Electronic Projects to Control Your Home Environment

Delton T. Horn

TAB Books
Division of McGraw-Hill, Inc.
New York San Francisco Washington, D.C. Auckland Bogotá
Caracas Lisbon London Madrid Mexico City Milan
Montreal New Delhi San Juan Singapore
Sydney Tokyo Toronto

Notices

Mylar E.I. Dupont de Nemours
Radio Shack Tandy Corp.

FIRST EDITION
FIRST PRINTING

© 1994 by **TAB Books**..
TAB Books is a division of McGraw-Hill, Inc.

Printed in the United States of America. All rights reserved. The publisher takes no responsibility for the use of any of the materials or methods described in this book, nor for the products thereof.

Library of Congress Cataloging-in-Publication Data

Horn, Delton T.
    Electronic projects to control your home environment / by Delton T. Horn.
      p.   cm.
    Includes index.
    ISBN 0-07-030416-5.   ISBN 0-07-030417-3 (pbk.)
    1. Electronic apparatus and appliances—Design and construction-
-Amateurs' manuals. 2. Detectors—Design and construction-
-Amateurs' manuals. 3. Environmental monitoring—Amateur's manuals.
4. Dwellings—Automation—Amateur's manuals. I. Title.
TK7870.H63 1994
681'.2—dc20   93-38580
                                                                                              CIP

Acquisitions editor: Roland S. Phelps
Editorial team: B.J. Peterson, Editor
                Susan Wahlman Kagey, Managing Editor
                Joanne Slike, Executive Editor
                Joann Woy, Indexer
Production team: Katherine G. Brown, Director
                    Patsy Harne, Layout
                    Kelly S. Christman, Proofreading
Design team: Jaclyn J. Boone, Designer
              Brian Allison, Associate Designer   EL2
Cover design: Carol Stickles   4437

Contents

Introduction *vii*

❖ **1 Sensing the real world electronically** *1*
Temperature sensors *2*
Light sensors *16*
Pressure sensors *24*
Position switches *28*
Other exotic sensors *29*
Homemade sensors *33*

❖ **2 Temperature-related projects** *34*
Heat-leak snooper *34*
Alternate heat-leak snooper *40*
Hot-spot locator *44*
Over/under temperature alert *49*
Simple electronic thermometer *52*
Voltmeter thermometer adapter *58*
Electronic thermostat *65*
Temperature equalizer *69*
Long-term thermometer *73*
Heat-activated fan controller *81*
Air-conditioner energy saver *84*
Power usage meter *93*

❖ **3 Liquid-related projects** *97*
Liquid sensor *98*
Moisture detector *102*
Simple plant-watering monitor *105*
Deluxe plant-watering monitor *110*
Flooding alarm *113*
Visual flooding alarm *116*
Sump-pump controller *119*
Water-heater controller *121*

❖ **4 Atmosphere-related projects** *125*
Wind-speed indicator *125*
Humidity meter *137*
Heater humidifier *148*
Air ionizer *152*

❖ 5 Light-related projects   *163*

Light-operated relay *164*
Dark-operated relay *165*
Self-activating night light *167*
Light dimmer *169*
Light cross-fader *171*
Sequential light controller *173*
Automated guest greeter *175*
Photosensitive automatic porch light *179*
Remote burned-out bulb indicator *182*

❖ 6 Technological risks   *186*

The problem of proof versus disproof *187*
Electromagnetic-field detector *191*
Radioactivity monitor *201*

Index   *213*

Projects

Project	Description	Page
1	Heat-leak snooper	*34*
2	Alternate heat-leak snooper	*40*
3	Hot-spot locator	*44*
4	Over/under temperature alert	*49*
5	Simple electronic thermometer	*52*
6	Voltmeter thermometer adaptor	*58*
7	Electronic thermostat	*65*
8	Temperature equalizer	*69*
9	Long-term thermometer	*73*
10	Heat-activated fan controller	*81*
11	Air-conditioner energy saver	*84*
12	Power usage meter	*93*
13	Liquid sensor	*98*
14	Moisture detector	*102*
15	Simple plant-watering monitor	*105*
16	Deluxe plant-watering monitor	*110*
17	Flooding alarm	*113*
18	Visual flooding alarm	*116*
19	Sump-pump controller	*119*
20	Water-heater controller	*121*
21	Wind-speed indicator	*125*
22	Humidity meter	*137*
23	Heater humidifier	*148*
24	Air ionizer	*152*
25	Light-operated relay	*164*
26	Dark-operated relay	*165*
27	Self-activating night light	*167*
28	Light dimmer	*169*
29	Light cross-fader	*171*
30	Sequential light controller	*173*
31	Automated guest greeter	*175*
32	Photosensitive automatic porch light	*179*
33	Remote burned-out bulb indicator	*182*
34	Electromagnetic-field detector	*191*
35	Radioactivity monitor	*201*

Introduction

This book is about environmental electronics. That does not mean it is about high-tech ecology. Instead, the subject is the design of electronic projects that interact with the environment in some way. As it happens, many of these projects do address ecological concerns of one type or another, and these concerns are discussed where appropriate.

Most electronic components are only "aware" of electrical parameters such as voltage or resistance, but humans rarely deal with such things directly. Humans are more interested in the features of the natural environment—things like light, temperature, moisture, and so forth. These are the kinds of things the projects in this book are designed to respond to.

Chapter 1 gives some basic background on commonly used sensor devices. The remaining chapters in this book present 35 projects that interact with the environment in some way. The chapters are organized according to what environmental feature the projects respond to.

Chapter 2 deals with temperature-sensing projects of many types. Several of the projects are designed to maximize the efficiency of your heating and cooling system, to save you money and help conserve energy—a major ecological concern.

Usually you should keep all electronic circuitry as far away from water (or other liquids) as possible, but the projects presented in chapter 3 are designed to monitor liquids. These projects sense when a container has been filled (or emptied) or when problems such as flooding or a slow leak occur. (Of course, only the actual sensors should be permitted to get wet. The rest of the circuitry should be kept as dry as possible, as is true for any electronic project.)

The projects in chapter 4 deal with the atmosphere in some way. Several projects monitor or enhance the humidity of the air in a room. This chapter also features an air-ionizer project.

Chapter 5 has several projects that respond to light in a variety of ways, for many different applications.

Finally, in chapter 6, I consider some ecological or environmental issues in a little more depth. There is much concern today about health risks from industrial sources. This chapter covers many of the current controversies over the risks of electromagnetic fields and radioactivity. A lot of unscientific nonsense has been published about these subjects, much of it verging on paranoia. Although it is impossible to discuss such complex subjects adequately in a single chapter, I have attempted to explain what the legitimate concerns are and what probably isn't worth worrying about. And you don't have to take my word for it. A simple but effective pair of projects for monitoring each of these modern environmental threats are featured.

I strongly encourage you to experiment with the circuits presented in this book. Some starting suggestions about possible areas of experimentation are included where appropriate. If you use a little imagination, I'm sure you'll be able to come up with some variations that haven't occurred to me. Environmental projects are particularly good for experimentation. In many cases, you can adapt a project to a totally different application, simply by substituting a different type of sensor.

I hope you enjoy these projects and find at least some of them useful.

❖1
Sensing the real world electronically

Usually electronic circuits are more or less self-contained, in their own specialized electrical world. To be useful, there will usually be some sort of input device—such as a manual switch, knob, or push button—and some sort of output device that converts the output of the circuit into some form humans can use—usually a visual or audible output. But usually, an electronic circuit is completely oblivious to the outside world. Electronic components are usually only "aware" of electrical parameters such as voltage, resistance, frequency, and so forth.

But in some cases, you might want an electronic circuit to interact more directly to the real-world environment. That is what you can do in all of the projects presented in this book. To do this, you need some special electronic components known as *sensors*. A sensor is sensitive to some environmental parameter, such as temperature, light, or moisture, and converts it into a proportionate electrical parameter, that can be dealt with by the rest of the circuitry.

In this chapter you can look at some basic types of environmental sensors used in modern electronics. In the following chapters you use many, but not all, these sensors in the projects. For your general information some common sensor types not used in any of the projects are discussed.

It is recommended that you experiment with any electronic project. Most electronic circuits can be adapted to very different purposes with just minor changes in the circuitry, and some imagination on your part. An environmentally responsive project can often be adapted to an entirely different application, just by changing the sensor used as the input to the circuit. For example, imagine a circuit that senses temperature with a thermistor, and gives an appropriate reading on a panel meter. This electronic thermometer can be converted into an electronic light meter, simply by replacing the thermistor with a photoresistor. (Both of these sensor devices are discussed in this chapter.)

Even though all the sensor devices discussed in this chapter are not used in the projects, you can still use them in your own modified version of one or more of my projects, designed to suit your applications.

Always use your imagination. Whenever you encounter a new electronic circuit, try to think of at least one or two ways you could adapt it for some application the original designer didn't intend, or perhaps didn't even consider. You'll get a lot more mileage out of any book of electronic projects that way.

Temperature sensors

Any practical electronic component is somewhat sensitive to temperature to some degree. Temperature changes will tend to change one or more of its operating parameters with changes in temperature. For example, the resistance of a component often changes with changes in temperature. This effect can be in either direction. If the resistance of the component increases as the temperature is raised, the device is said to have a *positive temperature coefficient*. Conversely, if the resistance of the component decreases as the temperature is increased, the device is said to have a *negative temperature coefficient*. Capacitance is another common electrical parameter that is frequently temperature sensitive.

Most *passive* components (resistors, capacitors, coils, etc.) have fairly small temperature coefficients. That is, a fairly large temperature change is required to make a noticeable difference in the electrical parameter of interest. Often normal, narrow-range temperature fluctuations are masked by the tolerance of the component. As an example, imagine a typical resistor, with a nominal value of 10 kΩ (10 kilohms or 10,000 ohms), a tolerance of 10%, and a positive temperature coefficient of 0.75 Ω per degree F (Fahrenheit). Assume the reference temperature is 70°F.

First, ignoring the temperature coefficient, the component's rated tolerance indicates that its performance will be considered satisfactory if its actual resistance is not more than plus or minus 10% of the nominal value. Because this is a 10 kΩ resistor, the actual value can be as much as 1 kΩ (1000 Ω) too high or too low. That is, the actual resistance can be anything between 9 kΩ (9000

Ω) and 11 kΩ (11,000 Ω), and it would still be considered acceptable as a 10 kΩ 10% resistor. Assume the particular resistor you are using has an actual resistance of 10,728 Ω at a temperature of 72°F. Because this component has a positive temperature coefficient, you can logically assume that if the temperature is increased enough, the actual resistance will be forced out of the acceptable tolerance range—above 11,000 Ω in this example. The temperature must be increased enough to raise the resistance by at least 272 Ω. You have already stated that the resistance increases 0.75 Ω for each degree of increased temperature. This means, the resistor in our example will be thrown out of tolerance only when the temperature is raised at least 363 degrees above the nominal room temperature of 72°F. This resistor will perform acceptably at temperatures up to about 435°F. This is a very good resistor, and you probably wouldn't have to worry about its temperature sensitivity at all in almost any normal application. If you did have to deal with temperatures this high, you'd probably have more problems keeping the solder and wire insulation from melting than with the increased resistance.

Some components have higher temperature coefficients. Try the same example, but this time assume the component has a positive temperature coefficient of 5 Ω per degree F. In this case, a temperature increase of just 55° (for a total temperature of 127°F) will throw the component out of its acceptable tolerance range. This could be a real possibility for many practical applications. In this case, the component temperature coefficient would be of significant concern to the circuit designer. This particular example is a pretty lousy resistor for ordinary purposes. (But it would be a fair thermistor in other applications. Thermistors are discussed later in this chapter).

Semiconductor components are particularly temperature sensitive. The delicate crystalline structure of the semiconductor can easily be damaged by excessively high temperatures. But even in the range of temperatures that are not harmful, the number of carriers crossing the threshold of a PN junction can be significantly influenced by relatively small changes in temperature.

Ordinarily, such temperature sensitivity is highly undesirable. Critical semiconductor components are often protected with some sort of *heat sink*, which is a metallic plate or fin designed to conduct heat away from the body of the component. A heat sink has a fairly large surface area, permitting it to dissipate more of its collected heat energy into the surrounding air.

A temperature-sensitive component can affect the accuracy of an electronic circuit, often in unpredictable ways, because the circuit might be used in a variety of environments with widely differing temperatures.

Temperature sensitivity can sometimes be an even greater problem than just loss of circuit accuracy (loss of accuracy is more than bad enough). It can even cause the component or the entire circuit to self-destruct.

In practical electronics work, there is no such thing as a perfect conductor. (Superconductors, which now exist only under highly specialized laboratory conditions, are not covered in this book.) When a current passes through any practical component, some of the energy will be wasted and converted into heat. This waste or loss is why components in a circuit carrying high-power levels can become quite warm to the touch. Imagine a typical transistor with a negative temperature coefficient for its internal resistance (which is the usual state of affairs for most practical transistors). That is, as the temperature of the transistor is increased, its resistance drops. What effect does this have on the current flowing through the device. Remember, Ohm's law tells you that the current flowing through any component is equal to the applied voltage, divided by the resistance of the device. That is:

$$I = \frac{E}{R}$$

Assume that the voltage being applied across this transistor is a constant +10 Vdc (volts direct current). This value will not change. Also assume that the resistance of the transistor starts out as 5000 Ω. The current flowing through the transistor is therefore equal to:

$$I = \frac{10}{5000}$$
$$= 0.002 \text{ A (ampere)}$$
$$= 2 \text{ mA (milliamperes)}$$

This current flowing through the transistor will inevitably increase its temperature somewhat. Now, suppose this conductance heat is sufficient to cause the negative temperature coefficient of the transistor to drop its internal resistance to 4500 Ω. Because the applied voltage remains unchanged, Ohm's law insists that the current flowing through the device must change to:

$$I = \frac{10}{4500}$$
$$= 0.00222 \text{ A}$$
$$= 2.22 \text{ mA}$$

Decreasing the resistance causes the current flow to increase. Because there is more current flowing through the transistor than before, it obviously must be producing more wasted heat. This effect combines with the negative temperature coefficient of the component to decrease the resistance, which increases the current flow, which increases the heat, which decreases the resistance, which increases the current flow and the heat, and so on, and so on in an ever escalating cycle. At some point, the internal resistance will be reduced enough that it will attempt to draw more current than it can safely handle. This excessive current flow will damage or destroy the semiconductor crystal, especially near the PN junction, rendering the device quite useless. The transistor will have self-destructed. This phenomenon is called *thermal runaway*. Thermal runaway can be a serious problem in many (though certainly not all) electronic circuits unless specific design precautions are taken to prevent it. For example, a current-limiting resistor can be used in series with the temperature-sensitive semiconductor component. Whenever you have two resistances in series, they must always carry exactly the same amount of current. The difference in their resistances will show up in the differing amount of voltage dropped across each component. No matter how much the internal resistance of the transistor drops, it will never carry more current than what is flowing through the current-limiting resistor. This arrangement can prevent the problem of potential thermal runaway in many practical circuits. Other design techniques, which are somewhat more sophisticated, are used in many electronic circuits to prevent such temperature-sensitivity problems, but you don't need to get into the details here, as long as you understand the general concept.

In some applications, a dedicated component that is specifically designed to respond to the ambient temperature in some predictable way might be very desirable, or even essential, to accomplish the purpose at hand. A dedicated sensor device will generally do a better job than the rather crude makeshift temperature sensors you have read about so far. A number of dedicated temperature sensors are readily available to the modern electronics hobbyist.

The simplest type of practical, dedicated temperature sensor is the *thermocouple*. Although there are a few commercially manufactured thermocouples on the market (especially from surplus dealers), in most cases it will be less fuss and bother to simply build your own. There is nothing very complicated or difficult involved in the construction of a thermocouple. A thermocouple is really nothing more than a simple junction of two dissimilar metals. If such a bimetallic junction is heated, a voltage that is proportional to the temperature of the junction will be developed across the two connecting wires. This is called the *Seebeck effect*, and is created by the different work functions of the two metals used in the construction of the thermocouple. Naturally, the specific operating characteristics of a practical thermocouple will depend heavily on the inherent electrochemical qualities of the two particular metals used in the device.

A similar approach is used in most standard mechanical thermostats. In a standard thermostat, two dissimilar metals are placed back to back. The difference in the rates of expansion and contraction due to changes in the temperature causes the two-metal "sandwich" to bend back and forth slightly. This physical movement can be used to open or close appropriate switch contacts to permit the thermostat to control a heater or air conditioner.

When using existing electronic components as heat sensors, a key fact to remember is that semiconductor materials tend to be much more temperature sensitive than ordinary conductors or insulators. This suggests that simple semiconductor components (such as diodes and transistors) could be used as temperature sensors. For quick-and-dirty temperature detection, an ordinary semiconductor diode will make a fairly good temperature sensor. Germanium diodes tend to be a little more temperature sensitive than silicon diodes, but germanium diodes are becoming increasingly difficult for electronics hobbyists to find. Fortunately, a silicon diode can still have an adequate temperature response (for most purposes), even though it isn't quite as good as that of a germanium diode.

For most silicon diodes, the voltage drop across the device will typically vary about 1.25 mV (0.00125 V) for every degree Fahrenheit of temperature change. If you use the Celsius (or centigrade) temperature scale, the change in the voltage drop across a typical silicon diode will be about 2.24 mV (0.00224 V) per degree. Notice that the diode is the same in both of these examples. The voltage drop/temperature ratio is actually identical in each case—the only difference is in what you are calling one degree. One degree on the Fahrenheit scale is smaller than one degree on the Celsius scale. As you can see, it is always vitally important to keep track of which temperature scale is being used, or the results will be completely meaningless and incoherent. (In some scientific literature, you might come across references to the Kelvin temperature scale. One degree Kelvin is exactly equal to one degree Celsius. The only difference is that the Kelvin scale is referenced to absolute zero, rather than the freezing point of water.)

Without getting too heavily into semiconductor theory, the secret of semiconductor diode operation is the PN junction. There are two types of semiconductor materials—N-type and P-type. An *N-type* semiconductor has too many excess electrons. A *P-type* semiconductor has too many excess *holes* (or too few electrons). Either type of semiconductor is nothing special by itself. It just acts like a moderately good conductor, rather like the carbon used in standard resistors. But when the two types of semiconductor are used together, some very special and useful properties show up.

A typical PN junction (or semiconductor diode) is shown in Fig. 1-1. You don't need to get into the details of the physics involved here. Many good books on semiconductor theory are available, if you are interested. For your purposes, you just need to know that current (electrons) can flow across the PN junction much more easily in one direction than in the other. In other words, the polarity of the voltage applied across the diode is vitally important. When the diode is *forward biased*, many electrons will flow across the PN junction. The device will exhibit a low electrical resistance. But when you reverse the polarity of the applied voltage, the diode becomes *reverse biased*. An ideal diode would block all current flow when reverse biased, but any practical diode will let a few electrons sneak across the reverse-biased PN junction, but not very many. The reverse-biased diode has a very high electrical resistance.

If you reverse all polarities, you can speak of the "flow of holes" instead of the more standard flow of electrons. Both expressions amount to exactly the same thing. All that's changed is the terminology.

The temperature will affect the flow of electrons through a semiconductor material, and thus the actual resistance of the PN junction. In most (but not all) practical temperature sensor

8 Sensing the real world electronically

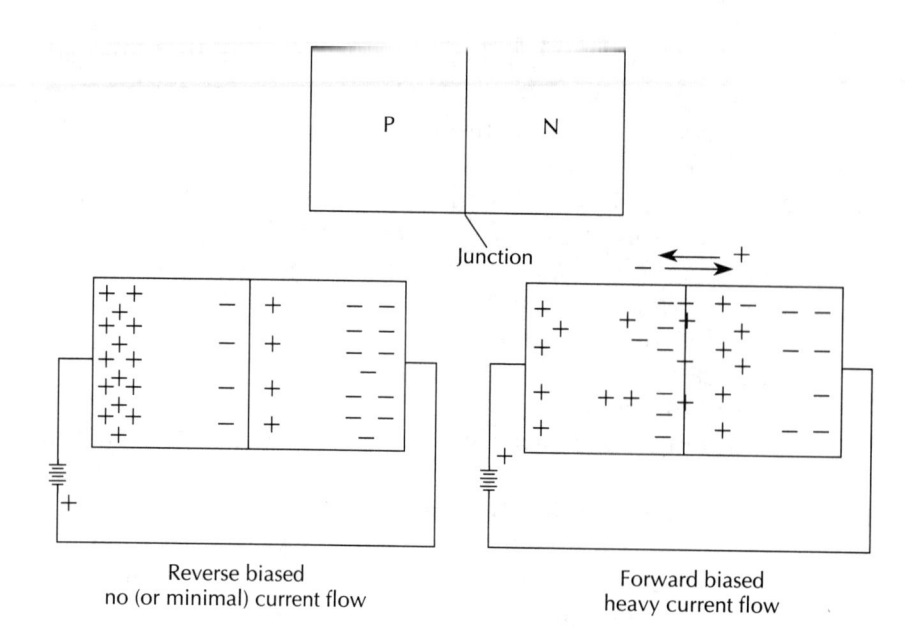

Fig. 1-1 *Typical PN junction or semiconductor diode.*

applications, the diode will be forward biased, but the effect works in either direction—it's just a question of which is easier to detect and measure in the circuit. The circuit is designed to detect differences in the resistance across the diode. The resistance of the diode can be measured directly, or, more commonly, the voltage drop across the diode will be measured instead. The voltage drop across any electronic component varies directly with its resistance, in accordance with Ohm's law.

A simple semiconductor diode can function fairly well as a temperature sensor, but it is rather crude. It is generally useful only a fairly limited range, and the response usually isn't very linear, which makes reasonably precise measurement over a range of temperatures very difficult to accomplish.

In many cases, a zener diode will make a better temperature sensor than a standard diode. Improved temperature response can also be obtained by using a transistor in place of a diode, as shown in Fig. 1-2. Notice that this is an NPN transistor wired as a diode, by shorting the base and collector leads together. In effect, the transistor now functions sort of like a super diode. In this configuration, the transistor base-emitter voltage is dependent on the collector current and the temperature. If you drive the collector with a constant current source, the only active variable will be the temperature, so you can determine the sensed temperature by measuring the transistor base-emitter voltage.

Fig. 1-2 *An improved temperature sensor can be made by using a transistor with a shorted base-collector connection instead of a simple diode.*

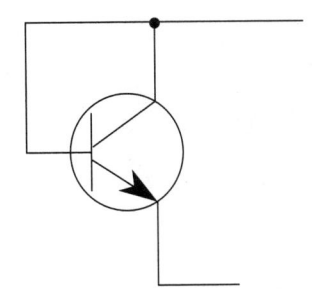

So far you have only been considering makeshift temperature sensors. You are using standard electronic components in ways they weren't really intended by their manufacturers. As is usually the case, you will get the best results by using a device designed specifically for the job at hand. There are several different types of dedicated electronic temperature sensors available to the modern electronics hobbyist. With a few relatively unimportant exceptions, they are all semiconductor devices.

The most common type of dedicated temperature sensor is the *thermistor*. The word *thermistor* is a contraction of thermal resistor, which describes the operation of the device quite well. It's popularity comes from the fact that it is inexpensive, accurate, and quite easy to use. Most commercially available thermistors offer a reasonably linear response over a fairly wide temperature range.

A thermistor is a junctionless semiconductor component. It has no PN junction, so it has no polarity. It can be wired into the circuit in either direction. It is designed to vary its resistance in response to the sensed temperature in a very predictable and consistent way. It is a two-terminal device. If the temperature is held constant, the resistance of the thermistor will not change. As far as the rest of the circuit is concerned, it is just an ordinary fixed resistor. For this reason, the standard schematic symbol for a thermistor is a variation on the standard resistor symbol, as shown in Fig. 1-3. Some sources omit the circle around the symbol because it does not add any information. However, the circle makes it easier to see at a glance that this is not just an ordinary resistor. But, the circle should always be used so the difference will stand out more.

Sometimes a small arrow will be added across the symbol, as shown in Fig. 1-4. This indicates that it is a variable resistance device. There is no practical difference in the component indicated if the arrow is included or omitted. It is a matter of personal preference. It is a very good idea to be as consistent as

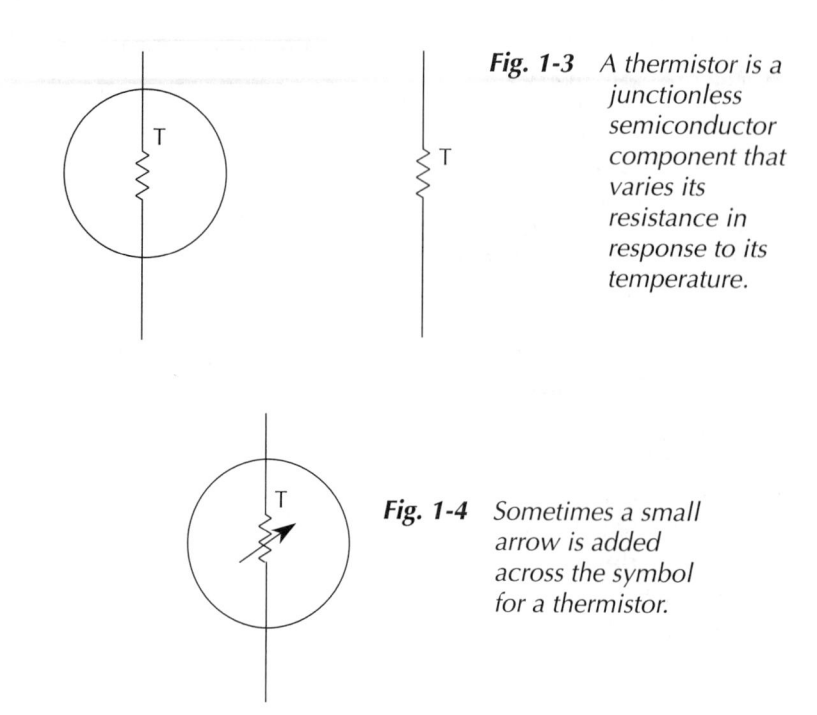

Fig. 1-3 A thermistor is a junctionless semiconductor component that varies its resistance in response to its temperature.

Fig. 1-4 Sometimes a small arrow is added across the symbol for a thermistor.

possible—make your choice about which symbol you are going to use in any schematics you draw, and stick to it. Don't risk confusion by using two versions of the symbol.

Whether or not the circle or the arrow is used in the schematic symbol, the important feature is the small T, which is always a part of the symbol for a thermistor. Without the T, you've just got an ordinary resistor. Including the arrow would make it a variable resistor, or *trimpot* (trimmer potentiometer). Adding the circle without the T would just be meaningless. (The T, of course, stands for temperature.) The placement of the T is not absolutely critical, but again consistency will help minimize the possibility for confusion. The thermistor symbol is usually shown vertically, as in Fig. 1-5 and Fig. 1-6. The standard placement of the T is to the upper right of the resistor symbol, as shown in these diagrams. There is less standardization when the thermistor symbol must be shown horizontally. Some sources keep the T to the upper right, regardless of the rotation of the main body of the symbol, as shown in Fig. 1-7. Most sources will keep the T in the same relative location with respect to the resistance symbol. The T itself might be rotated horizontally along with the rest of the symbol, as shown in Fig. 1-7. Others will rotate the T position, but still show the letter

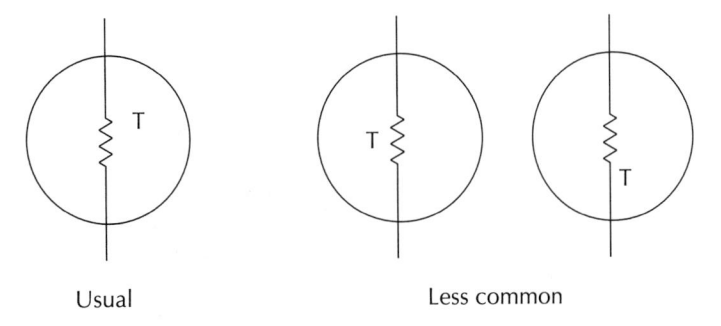

Fig. 1-5 The T in the schematic symbol for a thermistor is usually shown to the upper right.

Usual — Less common

Fig. 1-6 Some sources always keep the T to the upper right, regardless of the orientation of the main thermistor symbol.

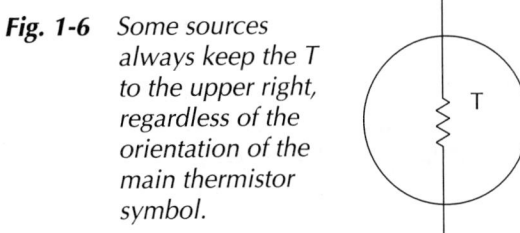

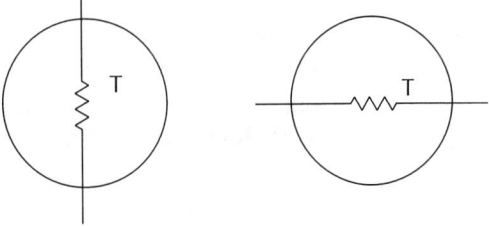

Fig. 1-7 When the schematic symbol for a thermistor must be shown horizontally, sometimes the T is rotated along with the rest of the symbol.

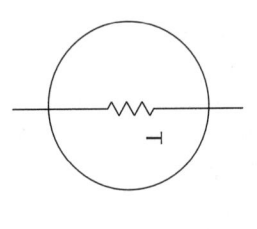

vertically, regardless of its location. This position is shown in Fig. 1-8. It doesn't make much difference really, as long as the *T* is clear and plainly visible, so the component can be correctly identified by someone reading the schematic diagram.

Most technicians try to draw thermistors vertically whenever it is at all possible to avoid the arbitrary choices and potential confusion involved in rotating the schematic symbol to the horizontal position.

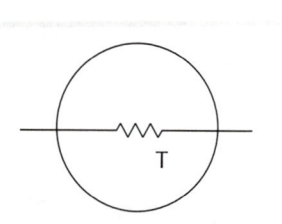

Fig. 1-8 *Some sources rotate the T with the rest of the thermistor symbol, but the letter itself is shown in its normal vertical position.*

There are two basic types of thermistors, defined by how they respond to changes in temperature. Just as with any other electronic component, a thermistor might have either a positive temperature coefficient or a negative temperature coefficient. A positive temperature coefficient thermistor increases its resistance as the sensed temperature rises. The hotter it is, the higher the temperature will be. The colder it is, the lower the resistance will be. A negative temperature coefficient thermistor behaves in exactly the opposite manner. The resistance of this component will decrease as the sensed temperature rises. The hotter it is, the lower the resistance will be, and the colder it is, the higher the resistance will be.

Negative temperature coefficient thermistors are, by far, more common than their positive temperature coefficient counterparts, although both types are available. In the projects presented in this book (and in most other similar literature), when an unspecified thermistor is called for in the parts list, a negative temperature coefficient device is normally assumed, unless otherwise indicated. In working with these devices, remember that the resistance across the thermistor decreases as it is heated. Cooling the thermistor increases its resistance. The thermistor resistance is at its maximum value at the lowest temperature of the monitored range and at its minimum value at the highest temperature encountered by the device.

Thermistors with positive temperature coefficients are sometimes called *sensistors* to avoid the awkwardness of the phrase *positive-coefficient thermistor*.

One problem with thermistors is that their response is not quite linear, making direct measurement techniques inefficient in electronic thermometer applications. This nonlinearity can be compensated for, but it requires a little advanced circuit trickery.

The nonlinearity of a thermistor will only be of significance if you are trying to precisely measure temperatures over a fairly wide range. Thermistors work very well in narrow range temperature-sensing applications. For example, they are ideal in any application that requires the circuitry to recognize and respond to a specific trigger temperature, or behave differently depending on whether or not the sensed temperature is above or below some critical value.

It is easy enough to alter the resistance range of a thermistor, simply by including it in series or parallel combinations with other resistive circuit elements. Assume you are using a thermistor with a resistance of 25 kΩ (25,000 Ω) at the highest temperature of interest, and 635 kΩ (635,000 Ω) at the lowest temperature of interest. If your circuit would work better with larger resistance values than this, you can add a range resistor in series with the thermistor, as shown in Fig. 1-9. This is just

Fig. 1-9 *In some circuits, it is desirable to add a range resistor in series with the thermistor. This will always increase the total resistance.*

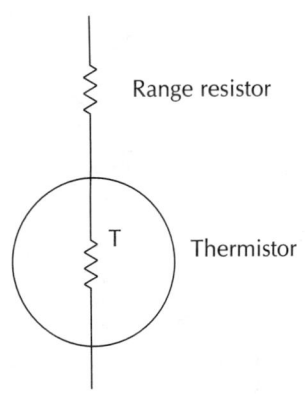

an ordinary fixed resistor. Assume it has a value of 470 kΩ (470,000 Ω). To determine the total resistance of resistances in series, simply add their values:

$$R_t = R_1 + R_2$$

At the highest end of the temperature range, the combined value will be:

$$R_t = 25,000 + 470,000$$
$$= 495,000 \; \Omega$$
$$= 495 \; k\Omega$$

At the opposite extreme, when the thermistor is at its coldest, the total effective resistance works out to:

$$R_t = 635,000 + 470,000$$
$$= 1,105,000 \; \Omega$$
$$= 1105 \; k\Omega$$
$$= 1.1 \; M\Omega$$

(Notice that the last value has been rounded—the extra 5 kΩ wouldn't be of much significance.)

On the other hand, use that same thermistor in a different circuit, with the same temperature range, but this time, the circuit would work better with lower resistance values than the thermistor produces directly. You can add a range resistor in parallel with the thermistor, as shown in Fig. 1-10, to lower the total effective resistance. Anytime there are two or more resistance elements in parallel, the total effective resistance will always be lower than any of the component resistances making up the combination. Once again, the range resistor is just an ordinary fixed resistor of an appropriate value. Use a 33 kΩ (33,000 Ω) range resistor in this example.

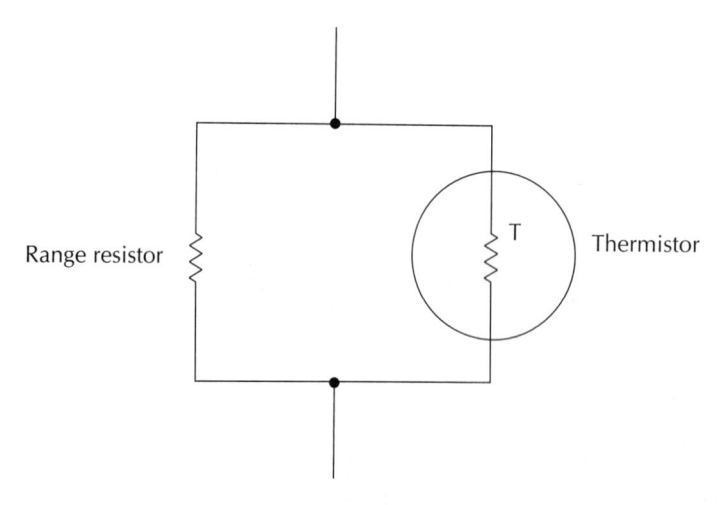

Fig. 1-10 *If a range resistor is placed in parallel with a thermistor, the total effective resistance will be reduced for any given temperature.*

There are two possible ways to calculate resistances in parallel. The general formula is:

$$\frac{1}{R_t} = \frac{1}{R_1} + \frac{1}{R_2} + \frac{1}{R_3} \ldots + \frac{1}{R_n}$$

This formula can be extended for any number of resistances in parallel.

The second method of calculating parallel resistance values will work only if there are only two resistances in parallel. This formula cannot be used if there are three or more resistance elements in the combination:

$$R_t = \frac{(R_1 \times R_2)}{(R_1 + R_2)}$$

Solving for the above example at the lowest temperature of interest, you get a total effective resistance of about:

$$R_t = \frac{(635,000 \times 3000)}{(635,000 + 33,000)}$$
$$= \frac{20,955,000,000}{668,000}$$
$$= \frac{20,955,000}{668}$$
$$= 31,370 \; \Omega$$

And, at the opposite extreme, when the thermistor is at its warmest, the total effective resistance works out to:

$$R_t = \frac{(25,000 \times 33,000)}{(25,000 + 33,000)}$$
$$= \frac{825,000,000}{58,000}$$
$$= \frac{825,000}{58}$$
$$= 14,224 \; \Omega$$

Notice that you have reduced the 610,000 Ω range (25 kΩ to 635 kΩ) to a range that runs just a little over 17 kΩ (14.2 kΩ to 31.4 kΩ).

In some practical applications, it might be desirable to use both a series range resistor and a parallel range resistor. A typical example of this type is shown in Fig. 1-11.

Fig. 1-11 In some practical applications, it may be desirable to use both a series range resistor and a parallel range resistor.

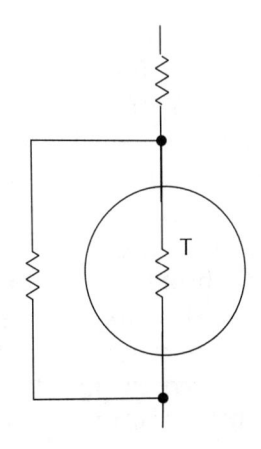

Dedicated temperature sensors ICs (integrated circuits) are now available from a number of manufacturers. There is considerable variation among such devices, of course, depending on the specific circuit design used in the chip, and the specific features and functions supported by the device. Most temperature sensor ICs are rather easy to use. Most have just three leads. They generally look like a slightly oversized transistor, or a zener diode with an extra lead added to its body. The three leads are used for the input voltage (V+, or just +), the common ground or negative voltage connection (GND, V−, or just −), and an adjustment terminal (ADJ), which is used to externally control the overall range of the sensor. The adjustment lead might not be used in all circuits in all applications.

When using such a temperature sensor IC, check the manufacturer's specification sheet for the specific circuit requirements and range limits of that particular device.

Light sensors

Another important element of the environment is light. Many electronic circuits use photosensitive devices to respond to the presence or absence of light levels above a specific threshold. Other photosensitive sensors operate over a continuous range, permitting the circuit to sense and respond to the intensity of the light striking the sensor.

The prefix *photo* simply means related to light, so a photosensitive device is one that responds to light. *Photons* (packets of light energy) strike the surface of a photosensitive material, affecting one or more of its electrical parameters. Most practical photosensors convert light energy into either a current flow, voltage, or a resistance.

All semiconductors are, by nature, photosensitive. The arrangement of free electrons (or holes) is disturbed in a predictable way by the photons striking the surface of the semiconductor material. This is known as the *photoelectric effect*.

In most applications, the photosensitivity of semiconductors is undesirable. Can you imagine using an audio amplifier that changed its output power and frequency response every time the lighting level in the room changed even a little? Such a device would be highly impractical and almost useless. This is why semiconductor components (transistors, ICs, etc.) are normally enclosed in light-tight housings—usually in a container made of black plastic or in a metallic can. These secure housings also protect the delicate semiconductor crystal from other environmental parameters, such as moisture, and, to a lesser extent, temperature.

Of course, in certain applications you do want to use the photoelectric effect. A wide variety of intentionally (and predictably) photosensitive semiconductor devices are available for the electronics experimenter to choose from. Following is a brief look at just a few of the most important and frequently used photosensors.

One of the handiest and simplest light-sensing devices available to hobbyist is the *photovoltaic solar cell*. Because of the recent concern about energy sources, much attention has been paid to various ways to convert sunlight into electricity. This development, together with efforts in connection with the space program, has led to the production of many low-priced solar cells.

When sufficiently bright light shines on the surface of a photovoltaic cell, a small dc voltage will be generated in the cell. The current flow will be dependent on the intensity of the detected light and the physical dimensions of the photovoltaic cell light-sensing surface.

Although photovoltaic cells are commonly referred to as *solar cells*, there is nothing at all to restrict their use only to sunlight. They will work with any light source. Of course, there is no energy conservation involved when an artificial (electrical) light source is used to power photovoltaic cells. The artificial light source must be consuming energy in some form in order for it to work at all. The well-established physical Laws of Conservation insist that you can never get more mass or energy out of a system than you put into it. This means the power generated, in a photovoltaic cell illuminated by the artificial light source, must always be less than the power used to operate the light source.

Photovoltaic cells are also popularly called *solar batteries*. Strictly speaking, it is not actually a battery unless there are two or more cells operating in series, or (somewhat less commonly) in parallel. But standard AA and D cells are commonly called batteries, even though they technically are just cells, not batteries. This rather informal and inexact terminology seems to have taken hold in the case of photovoltaic cells too.

Figure 1-12 shows a sketch of a typical selenium solar cell (the kind most readily available to the electronics hobbyist). Many different sizes are now available, but the standard is a little more than one inch square. The semiconductor material is spread out into a very thin sheet, applied to some sort of sup-

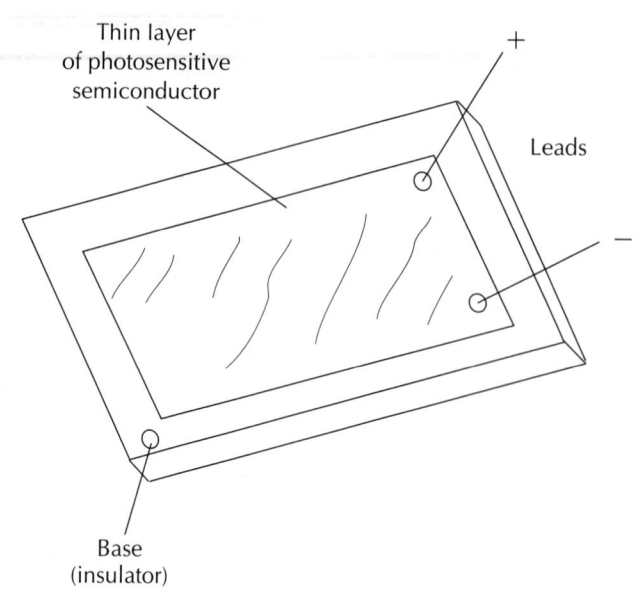

Fig. 1-12 *This is a typical selenium photovoltaic or solar cell.*

porting or backing material. Two flexible wire leads are connected to opposing ends of the wafer-thin semiconductor sheet. Some sort of protective lens is usually placed over the top of the cell to protect the semiconductor material from environmental damage. The side of the semiconductor exposed through the lens is photosensitive. The photovoltaic cell is fairly rugged and provides a large enough output that it is not particularly sensitive to noise effects. Although it is reasonably sturdy, a photovoltaic cell can easily be damaged or broken by rough handling or physical shocks. Never try to force a photovoltaic cell into a small space, where the cell might be bent. The odds are good that the semiconductor sheet will be cracked, either ruining the cell or significantly degrading its performance or reliability.

The photovoltaic cell produces an output voltage of about 0.5 V. This voltage is nearly constant, regardless of the amount of light reaching the cell and of the physical dimensions of the device. Once a little light reaches the photovoltaic cell, the output voltage will quickly rise to its final, maximum value. The output current of a photovoltaic cell, however, is directly proportional to the intensity of the light striking the semiconductor material. If the cell is only dimly lit, it will be able to supply only a very tiny amount of current. The higher the light intensity, the more current the photovoltaic cell can supply to any circuitry connected to it.

The photovoltaic cell output current is also affected by the physical dimensions of the photosensitive surface. The larger the semiconductor sheet, the more light energy it will be able to take in for a given source intensity, and therefore the greater the amount of current it can supply to a connected circuit.

The 0.5 V generated by a photovoltaic cell isn't very much, of course, and it is too low a voltage to be truly useful in most practical applications. Fortunately, this problem is easily overcome by using multiple photovoltaic cells in series. Voltage sources in series (assuming the polarities match) add. Three photovoltaic cells in series form a 1.5 V battery. A 3 V battery can be made up of six photovoltaic cells in series, and so forth.

Connecting photovoltaic cells in parallel increases their current-handling capability (again, the current capacity adds, assuming the polarities of the individual cells are consistent) but does not affect the voltage. Three identical photovoltaic cells connected in parallel will still only put out 0.5 V, but the available current will be three times that of a single cell.

Of course, the most obvious application for a photovoltaic cell is as a power source, just like an ordinary battery. This is certainly useful enough, but a photovoltaic cell can be used in many other practical applications as well. It can be used as a true sensor. The circuitry can detect the presence or absence of the voltage, or it might monitor and measure the current output, which will be proportional to the light level (a photovoltaic cell is basically a current generator). The photovoltaic cell is ideal for use with a transistor, which is essentially a current amplifier, or it can be used as a current-controlled electronic switch.

A photovoltaic cell (or other photosensitive device) can be used for object detection. For example, a circuit could be designed to count, or sound a tone, or otherwise respond each time a person passes through a doorway. This is done by shining a small light source directly on the photovoltaic cell, across the space to be monitored. When an object or person crosses between the light source and the photovoltaic cell, a shadow will be cast on the photosensor, which will respond to the change in lighting level. The circuitry should be designed to respond to this specific change in the photovoltaic cell's output. This type of system is shown in Fig. 1-13. This type of system can be used with most of the other photosensors you can read about in this section.

For such a system to work reliably, it will often be necessary to take steps to minimize the influence of ambient light in the area. The effect of ambient light can usually be limited by fo-

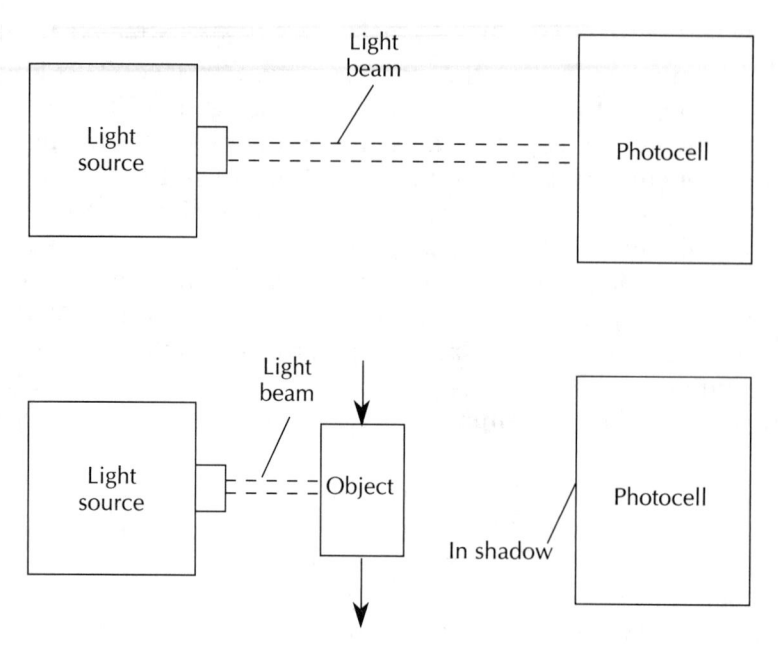

Fig. 1-13 *A photovoltaic cell can be used to monitor passing objects by shining a constant light source at the sensor across the path of the object.*

cusing the light source directly onto the surface of the photosensor and shielding the photovoltaic cell from all other light sources as much as possible. An easy way to do this is to place the source light and the photosensor each in a short length of tubing, as shown in Fig. 1-14. Often, cutting a cardboard tube from the center of a roll of toilet paper will be ideal. It often helps to paint the inside of the tube (especially the one shielding the photosensor) flat black.

If the light source is placed some distance from the photovoltaic cell, it might be necessary to focus the light beam so it doesn't diffuse too much before it reaches the photosensor, as shown in Fig. 1-15. The reflector from a flashlight will often do the job very inexpensively and conveniently.

The standard schematic symbol for a photovoltaic cell is shown in Fig. 1-16. Notice that it resembles the standard symbol for a regular single battery cell. The longer bar indicates the positive terminal. A small plus sign is also included to minimize the possibility of confusion. The incoming arrows indicate the light striking the surface of the component. Sometimes you will see this symbol drawn with two arrows, and sometimes with three. No functional difference is implied here. It's just a matter of the

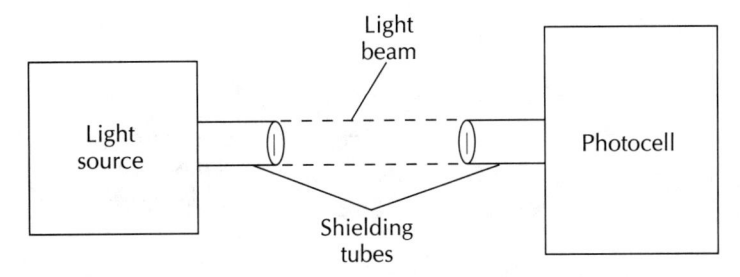

Fig. 1-14 Shielding tubes around the photosensor and the light source can help minimize unwanted effects from any ambient light in the area.

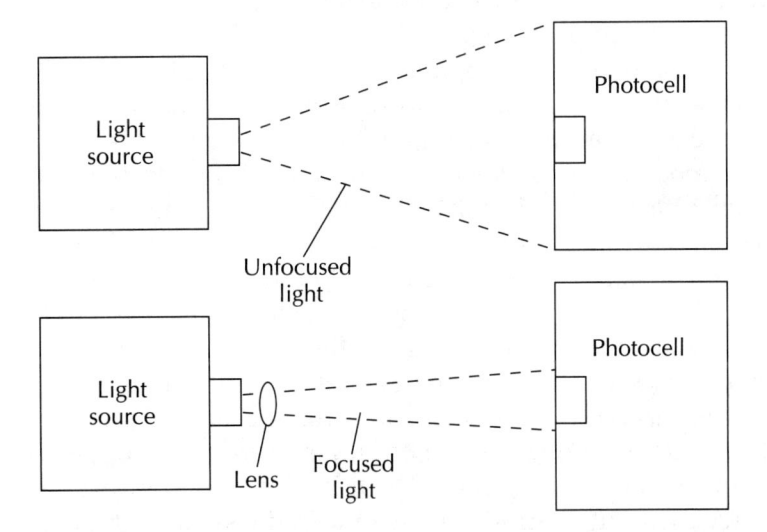

Fig. 1-15 If the distance between the light source and the photosensor is too great, the light might diffuse too much unless you use a focusing lens.

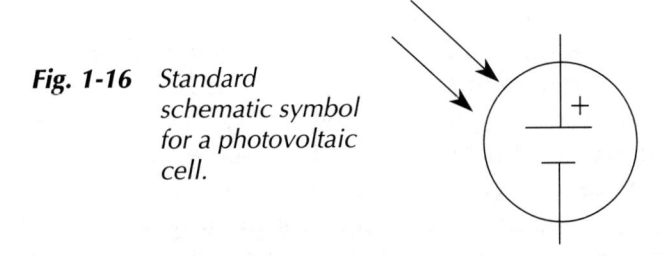

Fig. 1-16 Standard schematic symbol for a photovoltaic cell.

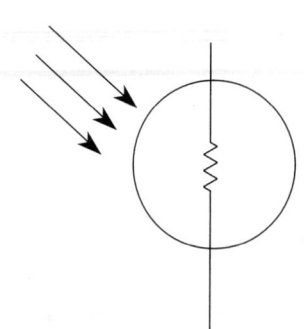

Fig. 1-17 Standard schematic symbol for a photoresistor.

individual preference of the technician drawing the diagram. However, showing only one arrow, or showing four or more arrows, would be incorrect and possibly confusing.

Another popular type of photosensor is the *photoresistor*. The name of this component pretty much says it all—it is a light-sensitive resistor. Occasionally, especially in some older technical literature, you might see this device referred to as a *light-dependent resistor* (LDR), or something similar. No functional difference is indicated by the alternate name. Today the term photoresistor is pretty well standardized.

The schematic symbol for a photoresistor is shown in Fig. 1-17. Notice how similar this symbol is to the one used to indicate an ordinary resistor. Once again, the important elements in the symbol are the arrows, which indicate the incoming light. The arrows must always be pointed towards the main resistor symbol. They usually (though not always) come from the upper right of the symbol, as shown here. Again, two or three arrows might be shown, without altering the meaning of the symbol at all.

The circle surrounding the resistor symbol is generally considered optional. Some technicians choose to omit it because it adds no additional information to the symbol. The circle is preferred because it makes it very easy to see, even at a glance, that the indicated component is not just a standard, ordinary resistor. When the circle is used, the arrows are always shown outside the circle.

Notice that, unlike the photovoltaic cell and most semiconductor components, the photoresistor is a nonpolarized device. Its two terminals are not specifically positive or negative. It can be used in either direction, just like an ordinary resistor. It is impossible to wire a photoresistor into a circuit backwards.

Most commercially available photoresistors have an inverse light/resistance ratio. That is, increasing the intensity of light shining on the surface of the component decreases its resistance, and vice versa. The photoresistor will exhibit its maximum resistance when in complete darkness.

A photoresistor has many potential applications. Theoretically, it can be substituted for any standard resistor in any circuit (although this isn't always desirable or advisable—sometimes the exact resistance value is critical to the operation of the circuit and should not be made variable). Essentially, you can use a photoresistor almost anywhere that you might consider using a manual potentiometer.

A *photodiode* can be a useful sensor, especially in electronic switching applications. The schematic symbol for a photodiode is shown in Fig. 1-18. Most of what was discussed about the symbol for a photoresistor applies here as well. The basic symbol is the standard diode symbol. As in the earlier symbols for photosensitive components, the important elements here are the arrows, which indicate the incoming light. The arrows must always be pointed towards the main resistor symbol. If they point outwards from the symbol, the device indicated will be a light-emitting diode (LED). Again, the arrows usually (though not always) come from the upper right of the symbol, as shown here. Again, two or three arrows might be shown, without altering the meaning of the symbol at all. The circle surrounding the diode symbol is generally considered optional. If the circle is used, the arrows are always shown outside the circle itself.

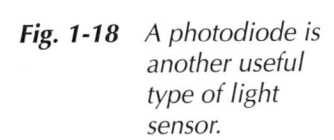

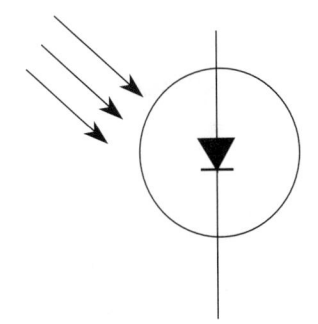

Fig. 1-18 *A photodiode is another useful type of light sensor.*

Another common semiconductor photosensor is the *phototransistor*. The schematic symbol for this device is shown in Fig. 1-19. This symbol is distinguished from that for a standard NPN transistor by the two or three incoming arrows, indicating photosensitivity. In this case, the circle is almost always included, unless the phototransistor is part of an optoisolator.

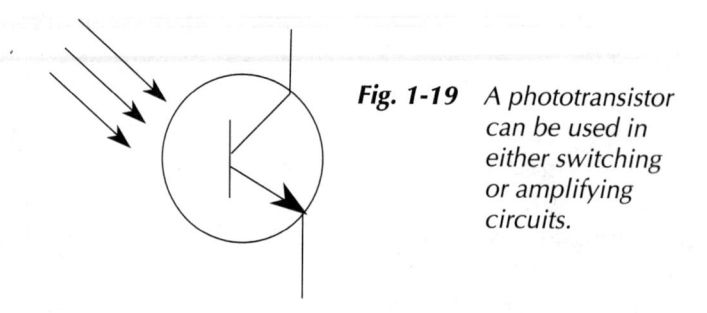

Fig. 1-19 A phototransistor can be used in either switching or amplifying circuits.

Usually the base lead of a phototransistor is omitted. The amount of light striking the sensor serves the same function as the signal applied to the base lead of an ordinary bipolar transistor. A few phototransistors do have a physical base lead, which might be used for external biasing, or left disconnected. Most phototransistors just have two leads—the emitter and the collector—which are used in exactly the same way as with an ordinary bipolar transistor.

Virtually all phototransistors today are of the NPN type, as shown here. PNP phototransistors do exist, but they are very rare and rather expensive. Apparently, there are special manufacturing difficulties involved in making PNP phototransistors.

Figure 1-20 shows the schematic symbol for a *LASCR* (light-activated silicon-controlled rectifier). It can be used in most low-power SCR applications. The light striking the sensor takes the place of the electrical gate signal in a standard SCR.

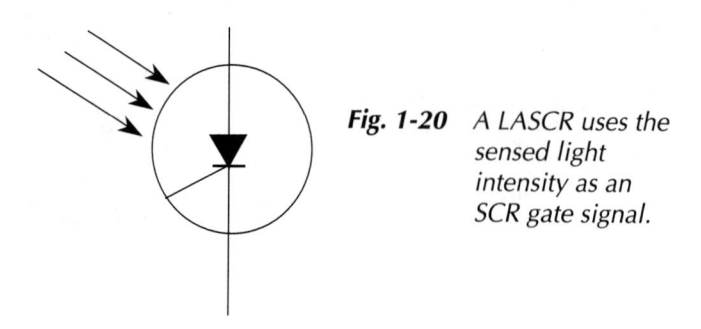

Fig. 1-20 A LASCR uses the sensed light intensity as an SCR gate signal.

Pressure sensors

Often it is helpful to convert mechanical energy into electrical energy. After all, the external environment is basically mechanical.

A *mechanical switch* is an obvious device for converting a mechanical movement into a form an electronic circuit can easily recognize and respond to. There is probably no need to go into switch theory here, because most likely you know how a switch works already.

Most switches are designed to be manually operated. You use your hand to move a slider or paddle from one position to another or to depress a push button. Obviously, this is mechanical energy. But sometimes it is helpful to have a switch that can respond to something other than your hand.

A pressure switch looks like a small mat. When sufficient pressure is placed on it, the switch is effectively closed. This is not a complicated device. Figure 1-21 shows the basic principle involved. A pair of flexible conductive sheets are separated by a small space, so there is no electrical connection between them. When pressure is placed on the upper sheet (the second sheet is assumed to be lying flat against a solid surface, such as a floor), it is distorted and bent downwards under the applied pressure. If the pressure is strong enough, the upper sheet physically touches the lower sheet, creating a short circuit between them—the switch is effectively closed.

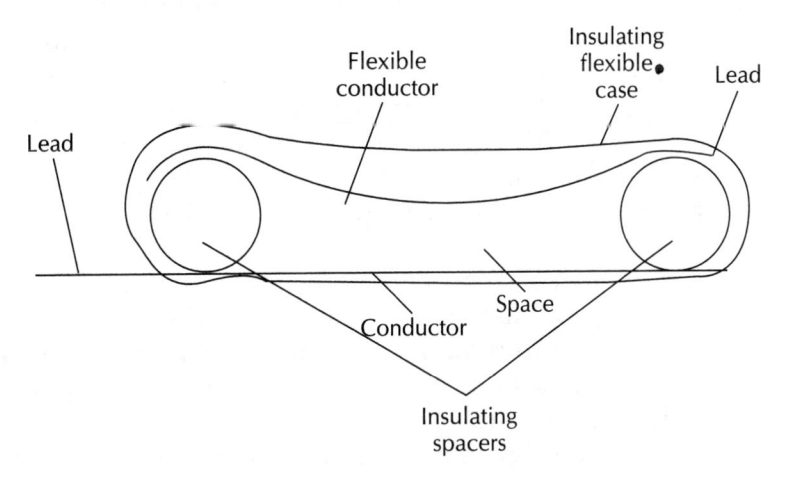

Fig. 1-21 *A pressure switch looks like a small mat. An internal switch is closed when sufficient pressure is placed on the mat.*

Snap-action switches are very useful for converting mechanical motion of various types into an electrical signal. This type of switch is shown in Fig. 1-22. A well-known switch of this type is the Micro Switch.

A *snap-action* switch is a tiny, momentary-action push-button switch, but you don't push the button directly. A long lever

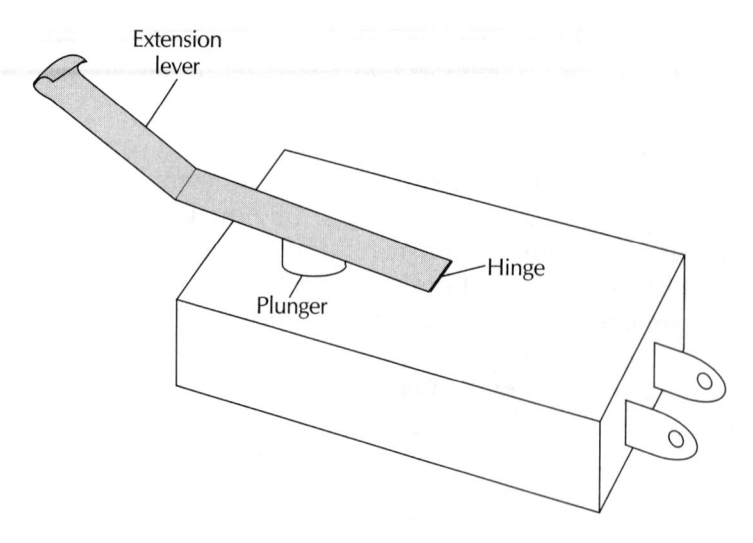

Fig. 1-22 *A snap-action switch, or Micro Switch can be used to convert very small mechanical motions of various types into an electrical signal.*

is extended out over the button. When the far end of the lever is mechanically moved (by whatever means), it forces the button down, activating the switch. Snap-acting switches might be either normally open (NO) or normally closed (NC). Most snap-action switches are SPST (single-pole, single-throw) units, but more complex switching functions such as SPDT (single-pole, double-throw) and DPDT (double-pole, double-throw) are also occasionally available.

Snap-action switches can be obtained with many different lever arrangements so they can be adapted to a number of different applications. As a rule, this type of device requires only a very small mechanical force or a very small physical displacement (or both) to activate the switch.

Several different snap-action switches can be placed so they will indicate different positions of a single object.

Air pressure switches are a little more exotic than the other mechanical sensor switches you have considered so far. But they can be very useful, especially for monitoring very weak mechanical pressure or movement. A typical air pressure switch can be closed by air pressures as low as 0.02 psig (pounds per square inch, gauge). This is about what might be felt from a gentle puff of air from a few inches away.

An air pressure switch can usually drive a LED of other low current circuit directly. To control circuitry with higher current requirements, the output of an air pressure switch can be used to control a relay or SCR, which electrically switches the higher power circuitry.

Sometimes a switch won't be adequate for the job you need done. A switch can only indicate simple, clear-cut yes/no conditions. In some applications, however, you might need to monitor pressure (mechanical energy) over a continuous range. This monitoring can be done by taking advantage of the *piezoelectric effect*, which occurs in crystals.

A crystal has two sets of axes running through it, as shown in Fig. 1-23. They are called the x axis and the y axis, and they are always perpendicular—that is, at a 90-degree angle from one another.

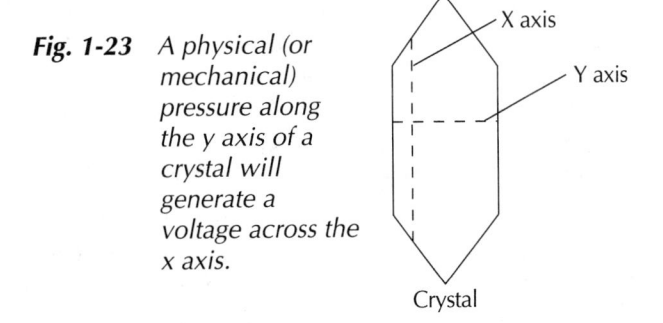

Fig. 1-23 *A physical (or mechanical) pressure along the y axis of a crystal will generate a voltage across the x axis.*

Physical (mechanical) pressure along the y axis of a crystal will generate a voltage across the x axis. The greater the pressure, the higher this voltage will be. Of course, this voltage can easily be detected and measured by an appropriate electronic circuit connected across the crystal.

Incidentally, although it isn't too relevant to your purpose here, it should be mentioned that the piezoelectric effect works in the opposite direction too. If an external voltage is applied across the x axis, a mechanical stress will be created along the y axis.

A variation on the basic piezoelectric principle is increasingly used in hybrid piezoresistive IC pressure transducers. These devices are found in many modern pressure-sensing applications in place of older mechanical pressure sensors.

These hybrid IC pressure sensors are smaller and more reliable than their mechanical crystal counterparts. Moreover, they are virtually insensitive to mechanical vibration and offer frequency response characteristics that allow operation right up through the audio frequencies.

Piezoresistive IC pressure transducers can be designed for a wide variety of pressure ranges, depending on the intended application. Units are available for measuring pressures from 0 to 5000 psig.

Mechanical pressure sensors are not always built around quartz crystals and the piezoelectric effect. A spiral coil of hollow glass, metal, or quartz can be used as a pressure sensor if one end is sealed. The coil will slightly wind tighter or unwind as the pressure of the liquid or gas (including ordinary air) within the hollow tube is varied.

Position switches

Mercury switches are useful for monitoring physical position, especially at an angle. A mercury switch is often called a *tilt switch*.

A typical mercury switch is shown in Fig. 1-24. Basically, this is just a sealed glass tube, containing two electrodes and a small globule of mercury. Leads are brought out from each of the electrodes to connect the device into a circuit. Mercury is a metal that is in liquid form at room temperature. This means the bit of mercury will flow freely about with the motion of the tube. Most of the time, the mercury will not be touching both electrodes, so there is no electrical connection between them. The switch is open. But if the mercury switch is tilted at the correct angle, gravity will pull the mercury down so it covers both electrodes. The mercury is conductive, like most metals,

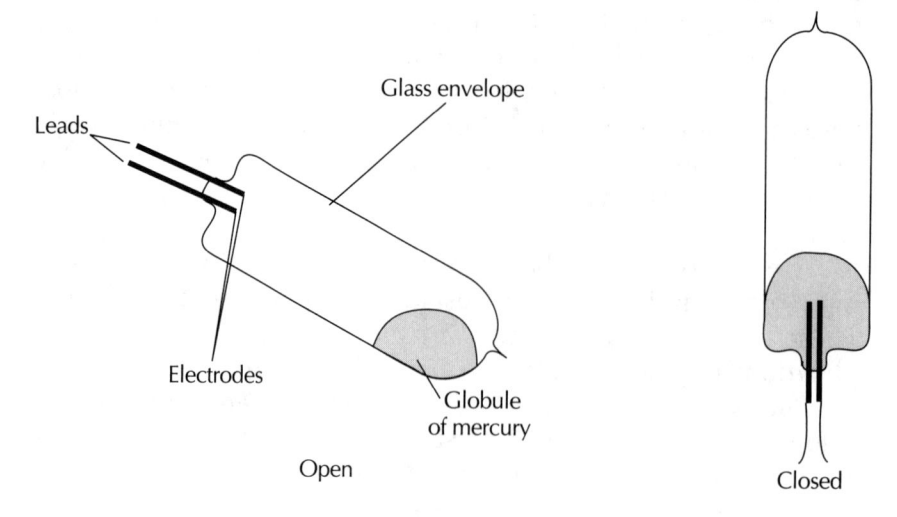

Fig. 1-24 *A mercury switch is often called a tilt switch.*

so a short circuit is formed between the electrodes. The current flows through the mercury to get from one electrode to the other. The switch is now closed. As you can see, the mercury switch is operated by its physical position or movement.

Mercury switches are always of the SPST, normally open type, due to the nature of their physical construction.

Magnetic reed switches are often used to monitor the position of mechanical objects, especially doors and windows. A magnetic reed switch consists of two physically separate parts. One part contains a small permanent magnet. No electrical connections are made to this unit. This part is usually mounted on the moving object itself. The other half of the magnetic reed switch contains the actual switch unit. This is a tiny reed switch. The contacts move physically when they are influenced by a magnetic field. When the magnet section is brought physically close to the switch section, the switch is activated. This action is shown in Fig. 1-25.

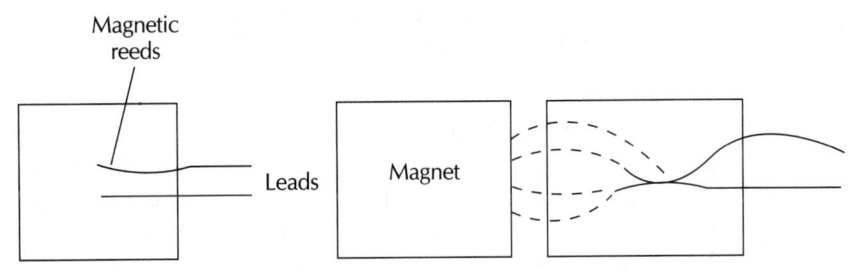

Fig. 1-25 *The contacts of a magnetic reed switch are designed to be moved when they are influenced by a nearby magnetic field.*

Most magnetic reed switches are SPST devices and are available in both normally open and normally closed versions.

Other exotic sensors

Today there are sensors available for almost anything you can imagine. Take a quick look at just a few of these rather exotic devices before closing this introductory chapter.

If a current flows through a conductor or semiconductor under the influence of a magnetic field at a right angle to the direction of the current flow (see Fig. 1-26), a voltage drop will be produced across the conductor (or semiconductor). In other words, its resistance will increase. This is known as the Hall effect.

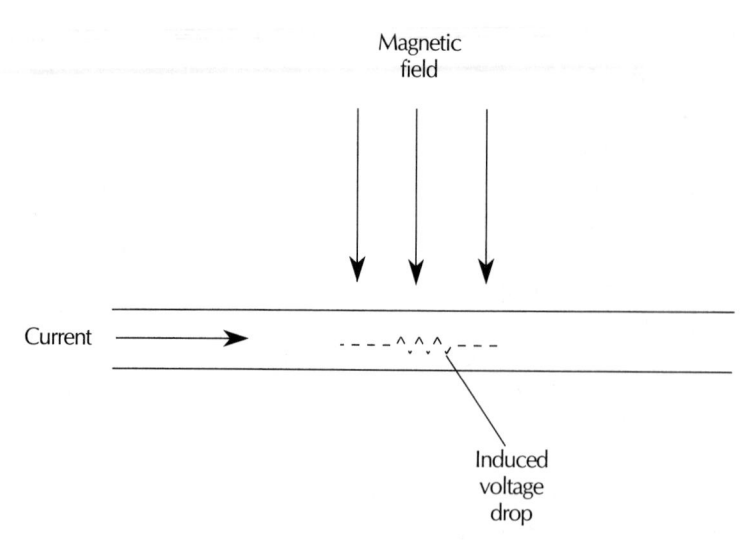

Fig. 1-26 *According to the Hall effect, if a current flows through a conductor under the influence of a magnetic field at a right angle to the direction of the current flow, a voltage drop will be produced across the conductor.*

Hall effect magnetic sensors can prove useful in a variety of monitoring, automation, or remote-control applications where a magnetic field of some sort exists. This application occurs more frequent than you might suspect at first. For example, any inductor (or coil or transformer) produces a magnetic field when current flows through it. This is also true of motors, because they include coils in their construction.

Sprague Electric Company is one of several manufacturers of Hall effect magnetic sensors in IC form. One such device is the UGN-3020T Hall effect switch, shown in Fig. 1-27. A block diagram of the internal circuitry is shown in Fig. 1-28.

To oversimplify matters a little, the amplifier stage boosts the output voltage from the actual on-chip Hall effect magnetic sensor. When the amplifier output signal level exceeds a specific threshold level, the Schmitt trigger turns on the output transistor.

Unwanted output oscillations are prevented by the built-in hysteresis of the Schmitt trigger stage. If the strength of the monitored magnetic field happens to be very close to the critical threshold level, ordinary noise effects might cause the detected field intensity to wobble back and forth on either side of the triggering point. Without hysteresis, the output would be er-

Fig. 1-27 *The Sprague UGN-3020T Hall effect switch can be a useful sensor device in certain specialized applications.*

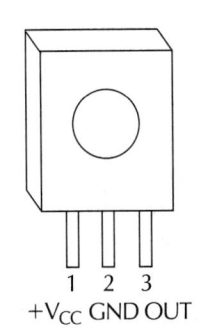

1   2   3
+V$_{CC}$ GND OUT

ratically switched on and off, confusing any circuitry controlled by the IC's output signal. The hysteresis built into the on-chip Schmitt trigger circuitry requires the amplifier output voltage to drop significantly below the turn-on level before the Schmitt trigger will snap back off. This greatly improves the overall stability and noise immunity of the device.

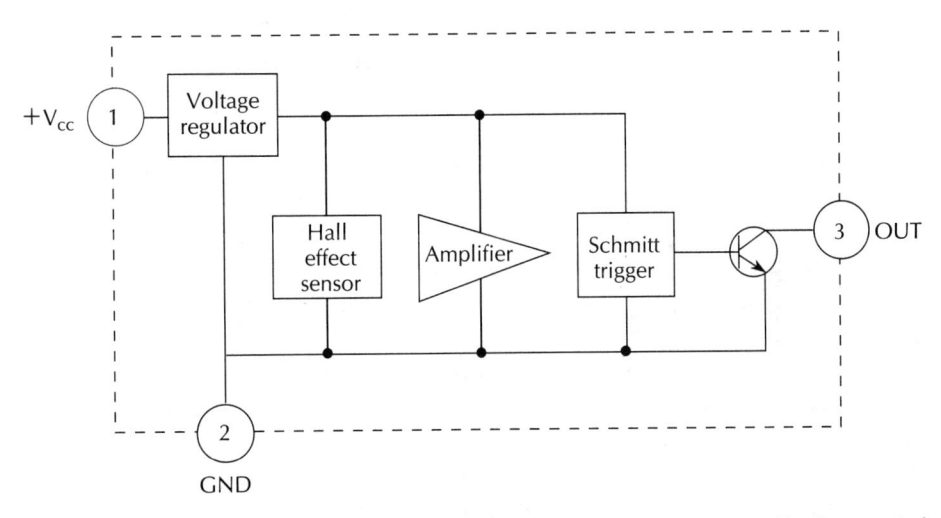

Fig. 1-28 *This is a block diagram of the Sprague UGN-3020T Hall effect switch circuitry.*

Electrochemical sensors also exist. They generally look for the presence of a specific chemical in the air, so they are usually called *gas sensors*, because they only respond to chemicals in gaseous form. You can loosely think of such a gas sensor as an *electronic nose*, because it "sniffs out" the chemical of interest.

Notice that most of the basic human senses can be roughly simulated (in very specialized ways) by electronic sensors. A gas sensor is an electronic nose, and a photosensor is an electronic eye. (In fact, in practical usage, photosensors are often called electric eyes by the general public.) Similarly, a microphone can serve as an electronic ear, and the sense of touch can be approximated by a pressure sensor. A taste sensor could serve as an electronic tongue, but are there practical applications for such a device?

Gas sensors are usually designed to detect the presence of poisonous or highly flammable gases, which could be very dangerous if permitted to accumulate undetected. Obvious applications for gas sensors would be in safety related equipment. They could set off an alarm when the dangerous gas is detected, or it could take some automated action, such as turning on an exhaust fan to ventilate the area, or close off a valve where the gas is presumedly escaping.

Other applications are also quite possible. For example, a gas sensor could be used to automatically adjust the fuel mixture in some sort of combustion engine or other device. Specific concentrations of the components of the mixture can be very precisely maintained this way, which would significantly improve the fuel burning efficiency of the engine.

Some specialized sensors can detect environmental attributes that human beings can't sense directly, but which could be harmful for a human being to be exposed to. An obvious example is radioactivity, which can only be detected by artificial means, such as an electronic Geiger counter. Radioactivity sensors are rather expensive and hard to find on the hobbyist level, but this is slowly changing due to the strong public concern (often bordering on paranoia) about the effects of radioactivity. Radioactivity sensors are usually rather delicate devices, and normally require rather complex and sophisticated support circuitry. Again, improvements are slowly appearing in this area.

Incidentally, there is a strong tendency among the general public to use the words *radiation* and *radioactivity* as synonyms. Light and heat both are forms of radiation, but not radioactivity. Radioactivity is a specific type of radiation involving the emission of high-energy subatomic particles. Another little known fact is that everything on earth (and probably all matter throughout the universe) is radioactive to some extent. The critical issue is how many high energy subatomic particles are emitted within a given period of time, and just how energetic they are (how fast they are moving).

The issue of radioactivity is far too complex to discuss in any greater detail here.

Homemade sensors

Sometimes you can build your own specialized sensor devices. For example a thermocouple (discussed above in the section on temperature sensors) is relatively easy to build. Using chapter 3 of this book, you can make your own moisture detectors for several of the projects.

If you have a special sensing application and can't find (or can't afford) a commercially manufactured sensor for the purpose, you can often design and build your own if you are imaginative enough.

In the discussions of the various projects presented in this book, tips are offered on home-brewed sensors and customization where appropriate. Remember, an environmentally sensitive project can often be adapted for a totally different application, simply by changing the sensor device providing environmental input to the circuit.

Be creative and have fun.

❖ 2
Temperature-related projects

An obvious element of any environment is the temperature. Is it warm, or is it cool? For human beings, some temperatures are comfortable, and others are not. Extremely cold or excessively hot temperatures can be dangerous or even fatal. Various types of electronic and mechanical equipment also have some degree of sensitivity to temperature. Any practical piece of equipment will have an ideal temperature range in which it can perform at peak efficiency. As the temperature moves away from this ideal range (in either direction), the equipment might operate less efficiently, or it might start to operate erratically or unreliably. It might even stop working altogether. In some cases, the equipment can be permanently damaged by attempting to operate it at extreme temperatures outside the acceptable temperature range of the device.

This chapter features a number of temperature-sensitive projects, many of which will help you save quite a bit on your utility bills.

Heat-leak snooper

By now everyone knows that insulation is a very important part of energy conservation. You obviously want the energy you use to heat or cool your home to stay inside where it can do some good. Insulation helps prevent, or at least limit, the unwanted energy transfer between inside and outside. But no practical insulation is perfect, of course. Few people live in sealed, air-tight chambers, so there will inevitably be some energy leaks in the insulation.

Energy leaks most commonly show up around the edges of windows and doors. Standard glass windowpanes also permit quite a bit of energy loss when the indoor and outdoor temperatures are significantly different. Other energy leaks are not so obvious. Often they are invisible to the eye. There are some energy leaks that can't really be explained. For example, in a heated room there might be a cold spot in the middle of a flat wall. Why is there an energy leak in the middle of what appears to be a solid wall? Perhaps an architect or building contractor could explain the hows and whys of such a situation. It happens, and it wastes energy.

It would not be practical to try to plug up every tiny energy leak in your home. It probably wouldn't even be possible. A few minor energy leaks will have only a negligible effect on the amount of energy required to heat or cool the building. But a severe energy leak can waste quite a bit of energy, and it should be plugged up or blocked in some way. But you have to find it first.

Finding energy leaks is the purpose of this project. The heat-leak snooper project is designed to detect temperature changes near an energy leak quickly.

For convenience in this discussion, assume that you are heating a room in winter. The temperature will drop near an energy leak. The project will work just as well for air conditioning in the summer—in this case, however, you will simply be looking for localized increases rather than decreases in temperature.

The heat-leak snooper project is very easy to use. It's output is a simple milliammeter. The position of the pointer indicates the relative temperature. In this project, the exact temperature is not of any particular importance, so it isn't even necessary to change the dial plate of the meter. In a perfectly insulated room, the meter pointer would remain stationary, regardless of where the sensor is placed.

The sensor for the heat-leak snooper is a simple thermistor in a hand-held probe. For maximum convenience, this project should be powered by a 9 V battery, rather than an ac-to-dc power supply. This way, you can carry it around the room with you, without worrying about getting tangled in or tripping over the power cord. The sensor should be mounted in a separate probe, connected to the main body of the circuit with a two-wire cable. Ordinary zip cord should work just fine for most applications. However, in some environments, there might be considerable rf (radio frequency) interference, which could affect the operation of the heat-leak snooper. If you happen to run into such problems, use a shielded two-conductor cable to connect the probe to the main circuit board. Such problems aren't too likely, but they are within the realm of possibility, especially if the project is used near heavy electrical equipment or radio transmitters.

Of course, the connecting cable should be flexible and reasonably strong so you can move the sensor around easily without risking breaking or unduly stressing one of the connecting wires.

The schematic diagram for the heat-leak snooper project is shown in Fig. 2-1. A suitable parts list for this circuit appears in Table 2-1. Because you are interested only in changes of temperature, rather than absolute calibrated measurement of the temperature, component tolerances are not too critical in this project. It

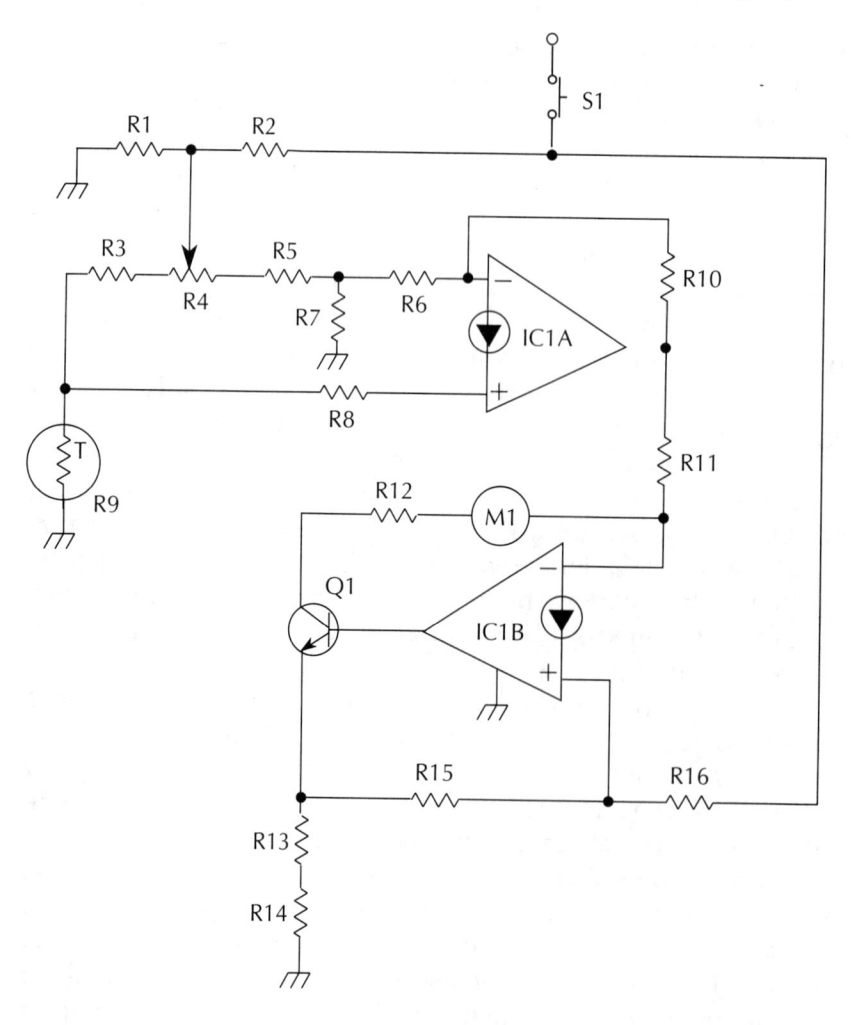

Fig. 2-1 Project 1. Heat-leak snooper.

Table 2-1 Suggested parts list for Project 1. Heat-leak snooper.

IC1	LM3900 quad Norton amplifier—see text
Q1	NPN transistor (2N5449 or similar)
M1	1 mA meter
S1	Normally open SPST push-button switch
R1, R15	10 kΩ, ¼ W 5% resistor
R2, R6, R8	22 kΩ ¼ W 5% resistor
R3, R5	2.2 kΩ ¼ W 5% resistor
R4	1 kΩ trimpot
R7	1 kΩ ¼ W 5% resistor
R9	Thermistor—see text
R10	2.2 MΩ, ¼ W 5% resistor
R11	8.2 kΩ, ¼ W 5% resistor
R12, R16	39 kΩ, ¼ W 5% resistor
R13	3.9 kΩ, ¼ W 5% resistor
R14	1.8 kΩ, ¼ W 5% resistor

is a good idea to use 5% tolerance resistors (rather than 10% or 20%), but there probably wouldn't be much practical advantage to using more expensive precision resistors in this circuit.

Look at IC1A and IC1B very carefully. Notice that these are not ordinary op amps (operational amplifiers). The schematic symbol for these devices is slightly modified to indicate the difference. Note the small arrowhead pointing from the inverting input to the noninverting input. This arrowhead indicates the device in question is a specialized variation of the basic op amp called the Norton amplifier. A standard op amp (even, an expensive high-grade unit) will not work in this circuit.

You don't need to go into the technical differences between the Norton amplifier and the standard op amp here. For your purposes, you just need to know that although the two devices are functionally similar, they are electrically incompatible, and one cannot be directly substituted for the other. They are similar, but different.

The LM3900 quad Norton amplifier IC called for in the parts list is relatively easy to find, and is generally quite inexpensive. It shouldn't cost you much more than a dollar. The LM3900 contains four Norton amplifier sections, but you are only using two in this project. You can leave the two unused sections idle. The LM3900 is cheap enough to waste half of it. Or you could build another project using two Norton amplifiers in the same housing as your heat-leak snooper.

Transistor Q1 isn't too critical. If you can't find the specific device suggested in the parts list, you should be able to substitute another low-to-medium power NPN transistor. Breadboard the circuit first to make sure the transistor you are using works well with the circuit. You might have to experiment with different resistor values if you substitute a different transistor for Q1. In this case, you should be especially concerned about resistor R12, and, to a slightly lesser extent, R13 through R15.

M1 is a standard 0 to 1 mA dc milliammeter. A larger meter will be easier to see, of course, but because you don't need to be too concerned about exact calibration markings in this project, you don't really need a particularly large dial face.

Notice that power is not continuously applied to this circuit. The supply voltage must pass through S1, a normally open SPST switch. This means the circuit will operate only when this button is held down. This helps prolong battery life. You only power up the circuit when you actually want to take a reading. If power were continuously applied, the battery would tend to wear down too rapidly. But if pushing the button is too much of a nuisance for you, you can replace S1 with an ordinary SPST slide or toggle switch. Just remember to turn the switch off when you are not actually using the heat-leak snooper.

The heat-leak snooper sensor is a thermistor (R9), which is a nonpolarized semiconductor component specifically designed to change its resistance in direct proportion to the temperature. It has a positive temperature coefficient, which means the resistance increases with increases in temperature. A number of thermistors are available in the electronics hobbyist market. Most have a positive temperature coefficient, but watch out. There are some negative temperature coefficient thermistors, and they would not be suitable for use in this project. Almost any positive temperature coefficient thermistor should work well with this circuit, although it might be necessary to adjust some of the resistor values in the input circuit for proper operation with some thermistors. If you do run into range problems, try experimenting with alternate values for resistor R8. In some cases, you might need to increase the value of trimpot R4 (while decreasing the values of resistors R3 and R5) to give a wider range of adjustment for calibration.

In this particular application, you scarcely need to worry about precise calibration. Many users will be satisfied with the project's performance without any calibration at all. But for best results, calibrate your circuit at room temperature. Simply adjust trimpot R4 for a midscale reading on the meter. This makes it very easy to notice any upward or downward movement of the pointer when looking for energy leaks, but it isn't really essential. If you're a stickler for exactness, you might want to use a ten-turn trimpot for R4, but it would be overkill for this type of application.

If you don't want to bother with calibration at all, you can replace R4 with two fixed resistors, as shown in Fig. 2-2. Probably the best choice would be two 470 Ω resistors, which would electrically be the equivalent to the trimpot set more or less at its midpoint.

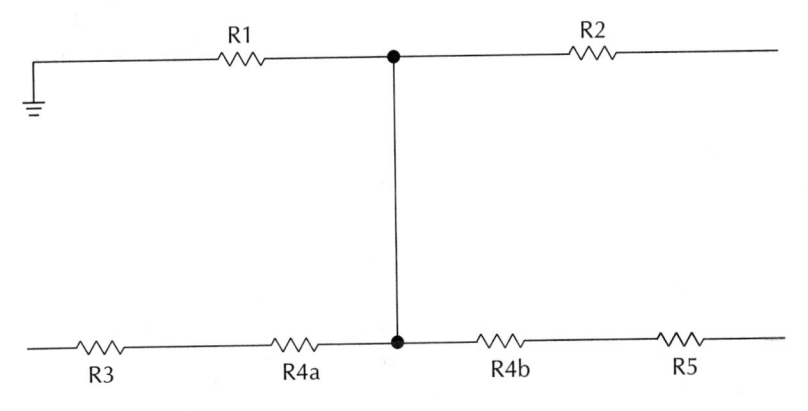

Fig. 2-2 *If you don't want to bother with calibration of your heat-leak snooper project, you can replace R4 with two fixed resistors.*

This heat-leak snooper project is quite simple and inexpensive, but it can be a big help in locating flaws in your home insulation, so you can correct them. The better your insulation is, the less energy you will waste, and the lower your utility bills will be. Who doesn't want that? You should be able to build this project for around $10, or maybe even less, and it shouldn't take more than an evening's work to construct the heat-leak snooper. Then, using the project to locate significant energy leaks throughout your home should only take an hour or two at most (depending on the size of your home and the number of energy leaks encountered). Over the next few years you could save literally hundreds of dollars on fuel with no decrease in your comfort. What a great bargain!

To use the detector, hold the main body of the instrument in one hand and the sensor in the other. Standing near the center of the heated room you want to check, press the activating button, and note the position of the meter pointer. Ideally, you should get the exact same reading at any point in the room. In practical terms, the reading will drop off slightly as you move farther from the heating vent(s). Walk around the room a little bit, keeping the sensor aimed away from any exterior walls, so you get a good feel of how much the temperature reading varies as you move toward or away from the heating vent(s). Notice that you will have to press the button each time you want to take a reading. This might seem like a bit of a nuisance, but it helps preserve the battery.

Now, slowly walk around the perimeter of the heated room, holding the probe near the exterior walls. Move it up and down to cover as much of the surface area as possible. Pay particular attention near any doors, windows, or other openings. Watch the meter carefully. If the reading begins to drop significantly when the sensor is at a specific point, you have located an energy leak that should be insulated.

This is one of those projects that is actually easier to use than to describe.

This project is probably one of the most important and generally useful projects in the entire book. If you are interested in building any of the other projects presented here, you will probably want to take advantage of this one.

Alternate heat-leak snooper

The function of the preceding project (the heat-leak snooper) is important enough that it makes sense to include an alternate version, achieving similar ends with different circuitry. This way, you won't be stuck if you run into problems locating parts for one version of the project. The schematic diagram for an alternate heat-leak snooper circuit is shown in Fig. 2-3. A suitable parts list for this project is given in Table 2-2. Notice that this heat leak snooper uses standard op amps, instead of the Norton amplifiers used in the previous version.

This project also uses a specialized IC, the LM335 temperature sensor, originally manufactured by National Semiconductor. As with any specialized electronics component, the marketplace is constantly changing, and there is no way of guaranteeing that it won't be obsolete and discontinued before you go to build this project. Make sure that you can locate the LM335s (the circuit uses two of these devices) before investing any money into the project to avoid possible disappointment and frustration.

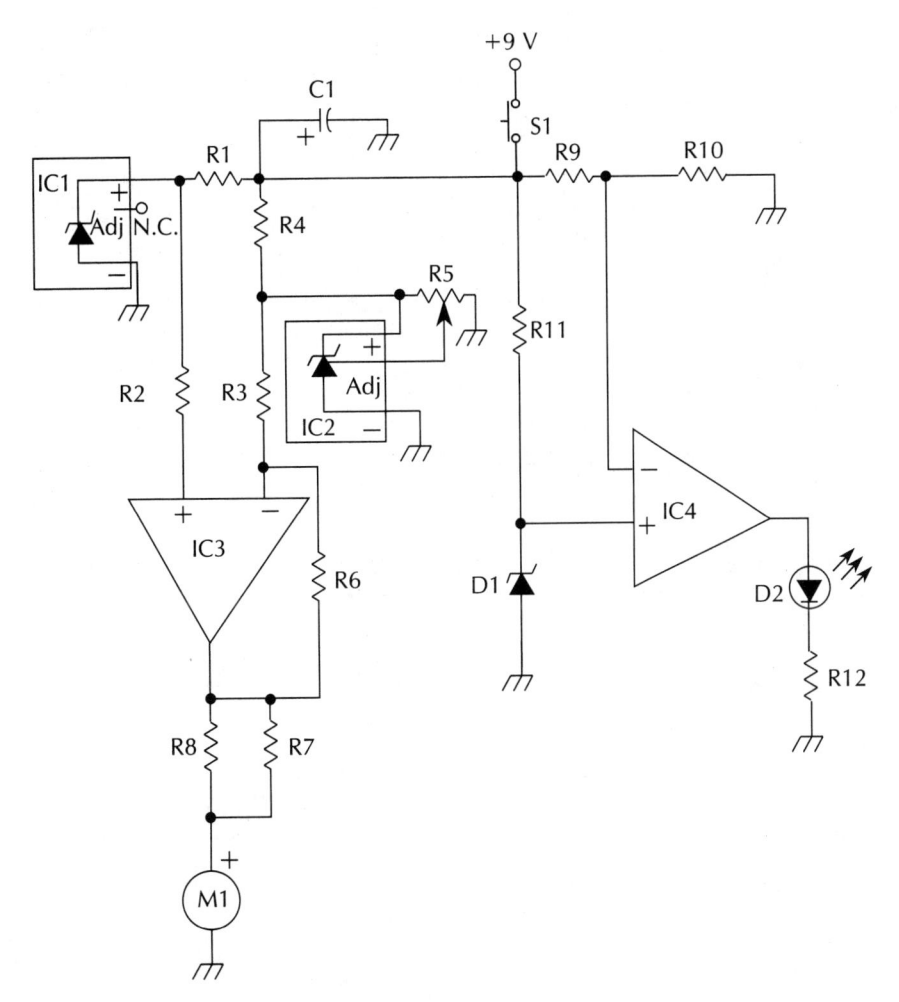

Fig. 2-3 *Project 2. Alternate heat-leak snooper.*

Several similar temperature sensor ICs have been developed by various semiconductor manufacturers. It should not be difficult for a moderately advanced electronics experimenter to modify the circuit to use some other component of this type. Consult the manufacturer's specification sheet for any substitute device very carefully to determine what modifications (if any) need to be made to the project circuit.

You shouldn't have any difficulty at all finding any of the other components called for in this circuit. They are all pretty standard and should be readily available from most electronics parts suppliers.

The pin-out diagram for the LM335 temperature sensor IC is shown in Fig. 2-4. Notice that this is a three-lead device, looking

Table 2-2 Suggested parts list for Project 2. Alternate heat-leak snooper.

IC1, IC2	LM335 temperature sensor
IC3, IC4	op amp
D1	5.1 V zener diode
D2	LED
M1	1 mA meter
S1	Normally open SPST push-button switch
C1	25 µF 35 V electrolytic capacitor
R1, R4	8.2 kΩ, ¼ W 5% resistor
R2, R3	15 kΩ ¼ W 5% resistor
R5	50 kΩ potentiometer
R6, R11	2.2 kΩ ¼ W 5% resistor
R7, R10	39 kΩ ¼ W 5% resistor
R8	270 kΩ ¼ W 5% resistor
R9	27 kΩ ¼ W 5% resistor

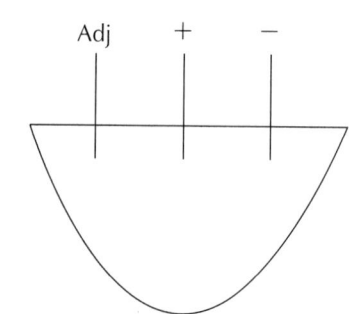

Fig. 2-4 *This is the pin-out diagram for the LM335 temperature-sensor IC.*

somewhat like a slightly over-sized transistor. The main leads are labelled + and −, indicating the dc polarity of the voltage applied to the device. The third lead on the LM335 is used to apply an external adjustment voltage for calibration and range-setting purposes. This adjustment pin is not used in all applications. In fact, in this project, you are using two LM335s (IC1 and IC2). One (IC2) uses the adjustment pin, and the other (IC1) leaves this lead unconnected.

The LM335 is much more accurate than the other temperature sensors used in other projects in this book (such as diode-wired transistors or thermistors). It has an accurate positive temperature coefficient of 10 millivolts per degree Celsius. It can be precisely calibrated to almost any desired temperature. It then functions over a range of 1°C.

Two LM335 temperature sensors are used in this heat-leak snooper circuit. IC1 is the actual functional sensor, mounted in an external probe. IC2 is a reference sensor, mounted in the main project housing. Do not pack the circuit too tightly near this component, and provide adequate ventilation so the reference sensor is not affected by ordinary heat build-up in the circuit. No electronic component is ever completely perfect. Some of the energy applied across any component is inevitably wasted, and this waste energy is converted into heat. You want the reference sensor to sense the general room temperature, not the operating temperature of its own circuitry.

In operation, the output signals from each of the two sensors (IC1 and IC2) are compared. Normally, they should be the same, but when the probe is near a heat leak, the difference in the detected temperatures will be indicated on the heat-leak snooper meter (M1).

Almost any standard op amps can be used in this project. High-grade, low-noise op amps might offer slightly improved performance, but the difference is likely to be negligible. You can use two separate op amp ICs (both of the same type), or a dual op amp, or half of a quad op amp package. As usual throughout this book, power supply connections and pin numbers are not shown for the op amps to avoid cluttering the schematic diagram, and to allow for the fact that different op amp devices might be connected differently. For example, some will operate off a single-ended power supply (V+ and ground), and others demand a dual polarity power supply (V+ and V−). When in doubt, check the manufacturer's specification sheet for the specific op amp device you intend to use in your project.

Calibration of this heat-leak snooper project is simple enough. Hold the project near the middle of the room, with the probe close to the body of the main instrument. (You don't have to get fanatic about this—just be sure the two sensors are reasonably close to one another.) This assures that both sensors should be at basically the same temperature. Now adjust potentiometer R5 for a midscale reading on the meter (M1). That's all there is to it. Notice that this calibration procedure automatically compensates for any moderate heat build-up within the main housing, which contains the reference sensor (IC2). Once calibrated, the two sensors will "think" they are seeing the exact same temperature. The temperature of the reference sensor (IC2) will presumedly remain constant. But if the probe sensor (IC1) is brought near an energy leak, its temperature will change, throwing the circuit out of balance. This change will be indicated by movement of the meter pointer from its calibrated midscale position. If the probe sensor is too cold, the pointer will move down the meter scale. If the probe sensor sees a higher temperature than the reference sensor, the pointer will move up the meter's scale. The greater the pointer movement, the more severe the detected temperature difference is. This project measures relative differences only. It would be difficult and impractical to calibrate the meter scale to read out the temperature difference directly. In normal applications for a heat-leak snooper, you won't need this information anyway. You just want to locate cold spots (or hot spots) within the room that indicate flaws in the room insulation.

To ensure the necessary portability, this project should obviously use battery power. Naturally, when the battery runs down too far, the circuit will not work accurately. To minimize such problems, a built-in low-battery indicator is included in the project as shown here. When the battery voltage drops too low, LED D2 will light up, alerting you to the need to replace the battery. If you don't want to bother with the low-battery alert function in your project, eliminate the following components from the circuit: IC4, D1, D2, R9, R10, R11, and R12. Eliminating these components will not affect the operation of the heat-leak snooper.

The heat-leak snooper (and the low-battery alert) will function only while pushbutton switch S1 is held closed. This helps minimize the drain on the battery. Push the button whenever you want to take a reading. The circuit will draw power from the battery only when it is needed. If you prefer, you can use an ordinary SPST toggle or slide switch in place of the normally open push-button switch recommended in the parts list. This will permit you to apply power to the heat-leak snooper circuitry continuously, without the (minor) nuisance of repeatedly pressing the button. Just remember to turn this switch off whenever you are not actually using your heat-leak snooper, or you will be wasting your battery power to no purpose.

The application and usage notes given for the preceding project also apply to this alternate heat-leak snooper project.

Hot-spot locator

The hot-spot locator project has a lot in common with the heat-leak snoopers presented earlier in this chapter. But the intended application is a little different here, calling for some changes in designing the circuitry.

The heat-leak snoopers were designed to locate relatively small increases or decreases from room temperature, indicating a flaw in the room's insulation. This project is intended to look for hot spots. It reacts to larger changes of temperature, and only in a single direction. It does not respond to cold spots.

This hot-spot locator project can also be used to spot leaks in insulation, especially in air-conditioned rooms. But it is more useful in servicing and troubleshooting. If a component, whether mechanical or electronic, isn't working efficiently, it will waste a lot of energy. This energy will be converted into heat, creating a hot spot. This circuit can be used to track down areas in a system that are significantly higher in temperature. This might be the source of an already existing problem, or it might be a source of potential future problems. This project can allow you to head off some equipment failures before they occur.

The hot-spot locator project can also be used to find hot spots in walls, indicating the presence of a fire or a potential fire. The usefulness of this application should be pretty obvious. One of the most common type of fires in homes is the electrical fire, which usually starts inside a wall where it can't be seen. By the time you are aware of the fire, it might be too late. A hot-spot locator can be used periodically to track down abnormally hot spots in the wiring, indicating damaged insulation, or some other potential problem that could soon lead to a tragic fire. You can correct the problem without incurring the losses of a fire. The necessary repairs will certainly be less expensive and easier before the fact too.

Unlike the earlier heat-leak snooper projects, this hot-spot locator project does not use a meter to indicate its findings. Instead, it gives an audible alert signal when it detects an abnormally hot spot. A small piezoelectric buzzer is used as the alarm-sounding device in this project. When the appropriate supply voltage is fed to this device, it produces a loud, unmistakable, and hard to ignore buzzing sound. The circuit is designed so current passes through the buzzer only when a hot spot is located. This project also features a visual output in the form of a LED (D1), which lights up when a hot spot is located.

The schematic diagram for this hot-spot locator project is shown in Fig. 2-5. A suitable parts list for this project appears as Table 2-3.

IC1 is a LM334 constant-current source. If you can't find this particular IC, you should be able to find some other suitable constant current source without too much trouble. You don't need to use a dedicated constant-current source IC. You

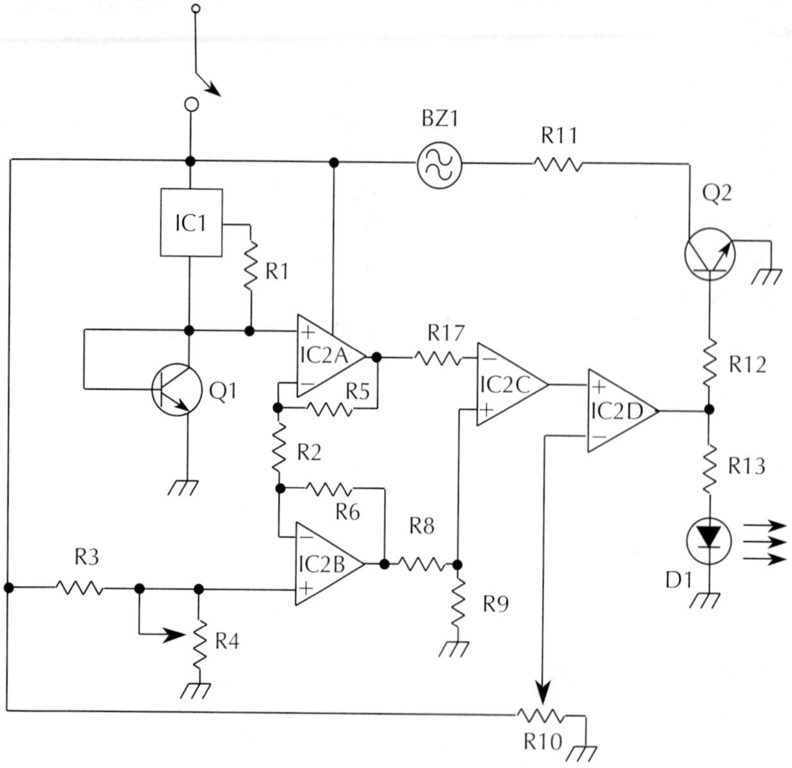

46 Temperature-related projects

Fig. 2-5 *Project 3. Hot-spot locator.*

Table 2-3 Suggested parts list for Project 3. Hot-spot locator.

IC1	LM334 constant current source
IC2	LM324 quad op amp
LED	D1
Q1, Q2	NPN transistor (2N3904 or similar)
BZ1	3 to 6 V buzzer—see text
R1	68 Ω ¼ W 5% resistor
R2	27 kΩ ¼ W 5% resistor
R3	15 kΩ ¼ W 5% resistor
R4	5 kΩ potentiometer—see text
R5, R6	150 kΩ ¼ W 5% resistor
R7, R8	100 kΩ ¼ W 5% resistor
R9	820 kΩ ¼ W 5% resistor
R10	10 kΩ trimpot
R11, R13	100 Ω ¼ W 5% resistor
R12	2.2 kΩ ¼ W 5% resistor

can construct a discrete constant-current source circuit around a transistor or a standard op amp. Such circuits are available in many technical books. In most cases, you won't need to make any changes in the circuitry to use such a substitution.

Except possibly for the LM334 constant-current source IC, this project calls for no unusual or specialized components, so you should have no real difficulty in locating all the parts you need.

The temperature sensor in this hot-spot locator circuit is an ordinary low-power NPN transistor, wired as a diode (with its base shorted to its collector). Remember, any semiconductor component is heat sensitive, which is ordinarily a disadvantage, but you can take advantage of this physical property in an application like this. Almost any low-power NPN transistor should work well enough in this circuit, but some devices might have slightly different operating characteristics, which might require some minor changes in some of the resistor values throughout the circuit. It would be a good idea to breadboard this circuit first and check it out thoroughly, before you heat up your soldering iron.

In most practical applications for this project, the sensor (transistor Q1) should be enclosed in a separate probe, rather than mounted on the main circuit board. To minimize possible interference problems, use a twisted-pair or a shielded two-conductor cable to connect the probe to the main circuit board.

Transistor Q2 is also not particularly critical in this circuit. Almost any low-power NPN transistor should work without modifications. This transistor functions as a simple amplifier and buffer for the output signal from op amp IC2D. There is no electrical reason requiring transistors Q1 and Q2 to be of the same type number, but this makes it a little more convenient to gather the parts for the project. After all, there is no reason to use different transistor type numbers for these two components either.

Almost any piezoelectric buzzer designed for operation on 3 to 6 V should work well. Some specific buzzers, and some transistors (Q2) might have slightly different characteristics, so you might find performance improved somewhat with a different value for resistor R11. Again, experiment with this component value while breadboarding this circuit. In some cases, you might be able to eliminate this resistor from the circuit altogether, although that is not recommended. A resistor acts as a current limiter in case of a short circuit or other problem that might develop in the circuit, which could protect some other components from potential damage due to current overload.

This project also features a visual output in the form of an

LED (D1), which lights up when a hot spot is located. Notice that the LED also operates in unison with the buzzer. If for some reason, you want an audio output indication only, you can simply eliminate resistor R13 and LED D1 from the circuit. This will not affect the operation of the project in any other way. Conversely, if you prefer to have only a visual indication when a hot spot is located, you can omit the buzzer (BZ1), resistor R11, and transistor Q2. However, using both forms of indication will make the project most convenient to use under a wide variety of practical conditions. Neither is particularly expensive to include in the project, so it wouldn't make much sense to eliminate one or the other output indication mode for economic reasons.

This circuit uses four standard op amp stages. Almost any standard op amps can be used in this project. All four should be of the same type. There might be some minor improvement in the circuit operation if you use high-grade low-noise op amps, but the improvement will probably be very slight and not really worth the increased expense.

You can use four separate single op amp chips, or two dual op amp ICs, but a quad op amp chip makes the most sense in this circuit. The LM324 is a good choice. It is inexpensive and widely available. Unlike many single op amp devices (like the popular 741), it can be operated from a single polarity power supply. A negative supply voltage is not needed.

Potentiometer R4 and potentiometer R10 are used to calibrate this hot-spot locator project. You can use two trimpots for this control. In some applications, it might be desirable to use a front-panel control for potentiometer R4. It would probably be overkill to use two front panel controls in this project. It would also make operating the device unnecessarily confusing.

The two controls interact somewhat. Potentiometer R13 controls the triggering of the final comparator stage (IC2D). It should be adjusted so the alarm does not sound except when the probe temperature is significantly different from the ambient temperature. How much temperature difference is significant? The significance pretty much depends on your specific application. This potentiometer gives you quite a bit of choice in the matter. For a wider operating range, trying using a trimpot with a larger full-scale resistance.

Potentiometer R4 is adjusted to calibrate for the current ambient temperature. This adjustment permits you to use the project in a variety of situations. A front panel calibration control for R4 would be particularly useful if the project is likely to be used outside, where you have little or no control over the ambient temperature.

Notice that there is no actual sensor to detect the ambient temperature in this project. Instead, you use a simple resistive voltage-divider network (made up of resistor R3 and potentiometer R4) to set up a reference voltage that corresponds to the ambient temperature as sensed by the sensor (Q1). Adjust R4 so that the voltage at the noninverting input of IC2B is the same as the voltage at the noninverting input of IC2A. Measuring the actual voltages is really only necessary during the initial calibration of the project. (It wouldn't hurt to periodically double check the calibration of any test equipment.) In practical operation, potentiometer R4 is adjusted so that the alarm does not sound while the probe is at or near the ambient temperature. When a significantly different temperature is sensed, the voltage comparator circuit is triggered and the alarm sounds. The LED (D1) lights up, indicating you have located a hot spot.

Over/under temperature alert

Sometimes, you don't really need to know the exact temperature so much as that it is within a specific range. That is, the temperature is neither too hot nor too cold. That is the purpose of the next project. This circuit will tell you when the monitored temperature goes beyond a preset range. It will even tell you if the out-of-range temperature is too hot or too cold.

The schematic diagram for the over/under temperature-alert circuit is shown in Fig. 2-6. A suitable parts list for this project appears in Table 2-4.

Basically, this circuit is a window comparator, which compares two voltages made up from a pair of resistive voltage-divider strings. One voltage divider (the reference) is made up of three fixed resistors (R1, R2, and R3). The other voltage divider (input) is comprised of a trimpot (R5) and a thermistor (R4). A thermistor, as you should recall, changes its resistance in direct proportion to its temperature. The trimpot is used to adjust the effective sensitivity of the thermistor, as far as the comparator circuitry is concerned.

There are three voltage comparator stages in this circuit. IC1A compares the temperature-dependent input voltage with the upper portion of the fixed reference voltage. At the same time, IC1B compares the temperature-dependent input voltage with the lower portion of the fixed reference voltage. Then, IC1C com-

50 Temperature-related projects

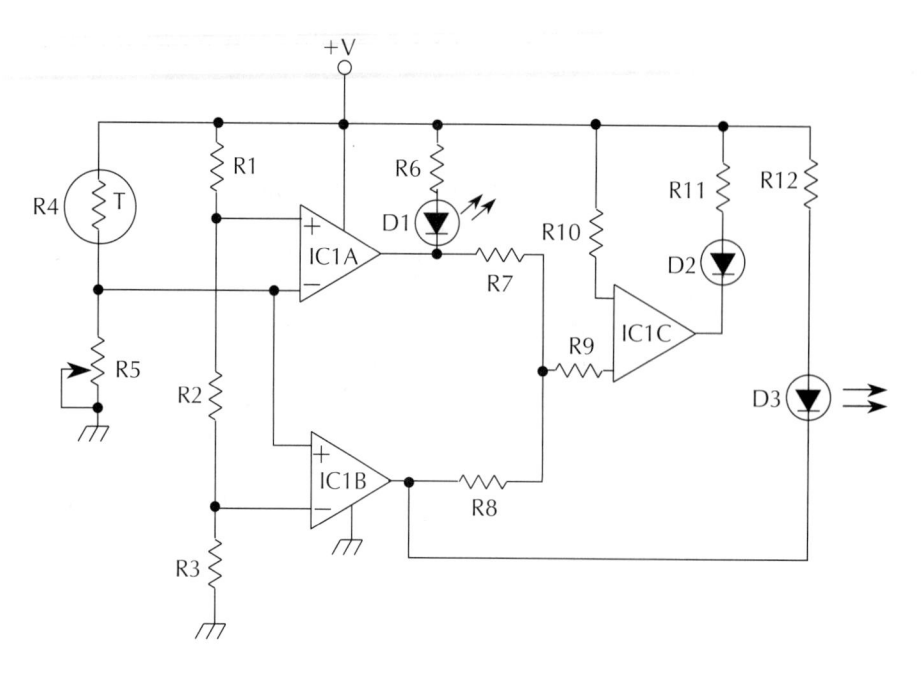

Fig. 2-6 *Project 4. Over/under temperature alert.*

Table 2-4 **Suggested parts list for Project 4. Over/under temperature alert.**

IC1	LM339 quad comparator—see text
D1, D2, D3	LED
R1, R3	22 kΩ ¼ W 5% resistor—see text (experiment to change switching points)
R2	10 kΩ ¼ W 5% resistor—see text (experiment to change switching points)
R4	Thermistor
R5	1 MΩ trimpot
R6, R11, R12	330 kΩ ¼ W 5% resistor
R7, R8	27 kΩ ¼ W 5% resistor
R9	47 kΩ ¼ W 5% resistor
R10	68 kΩ ¼ W 5% resistor

pares the combined outputs of the first two comparator stages with another reference voltage derived through resistor R10.

Three LEDs (D1, D2, and D3) are used to indicate the relationship of the monitored temperature to the preset range. Only one of these LEDs should be lit at a time. As long as power is applied to the circuit, one and only one of the LEDs should be lit at any given moment.

The individual meanings of these three LED indicators are as follows:

D1   temperature too high
D2   temperature in range
D3   temperature too low

Notice that the high/low voltages fed to these LEDs can easily be tapped off and fed to other circuitry within a larger system. They can feed data to a master computer, or they can activate heating or cooling devices when appropriate. There are many possible applications. Use your imagination.

This circuit is designed around three sections of an LM339 quad comparator IC. This is a fairly common and widely available chip. The LM339 contains four independent voltage comparator stages, but you only need three in this project. The fourth comparator section can be used in other circuitry, as part of a larger system, if that is appropriate to your intended application. If this extra comparator stage is left unused, however, its inputs and outputs should all be grounded. A comparator section with floating inputs or outputs can adversely affect the operation of the other comparator stages on the same chip. You would be liable to run into stability problems with the circuit.

The LM339 is designed specifically for voltage comparator applications, but you can substitute standard op amp chips if you prefer. Be aware that most standard op amps will require a dual-polarity power supply for this type of application.

The reference temperature range is set by the values of voltage divider resistors R1, R2, and R3, along with the specific characteristics of the particular thermistor used. Any standard thermistor should work in this circuit.

You will probably want to do some experimentation with alternate values for the resistors in the reference voltage divider string. They will set the too-high and too-low reference points in your project. The values of these three resistors interact, but, roughly speaking, you can define their functions, as follows:

R1   sets too-high temperature voltage
R2   sets spacing between too-high and too-low points
R3   sets too low temperature voltage

In some applications, it might be desirable to replace some or all of these resistors with trimpots, to permit manual calibration. For most purposes, it would probably be best to select a suitable fixed value for R2, then use trimpots for R1 and R3 to vary the too-high and too-low trip points of the comparator circuitry.

Remember, the values of these three resistances inevitably interact. Changing any one resistance alters the balance of the entire reference voltage string. For example, suppose you have carefully adjusted trimpot R1 to set the desired too-high temperature trip point. Then calibrate the too-low temperature trip point via trimpot R3. If you then go back and double check the too-high temperature trip point, you'll probably find it has changed, perhaps by a great amount, depending on how much you have changed the resistance of R3.

For many applications it will be more convenient to use "paper calibration." That is, calculate the appropriate resistances for your desired trip-point temperatures, then use fixed resistors in the circuit. What you loose in adjustability will be more than made up for in accuracy, and ease of use. Of course, it all depends on how necessary manual adjustments might be to your specific intended application for the project.

Simple electronic thermometer

A simple but effective electronic thermometer circuit is shown in Fig. 2-7. A suitable parts list for this project is given in Table 2-5. Although 5% tolerance resistors are suggested in the parts list, better results can be obtained if you use precision, low-tolerance resistors throughout the circuit.

It would also be a very good idea in most cases to use ten-turn precision trimpots for R6 and R7. Standard, inexpensive trimpots will give you a functional project, but it will be much more difficult to calibrate the thermometer accurately. If your application is not too critical, standard trimpots might permit you to get close enough to correct calibration. For example, if you are using the circuit as a simple room temperature thermometer, it probably won't matter too much if the read-out is a degree or two off from the true temperature value. However, if you are using the project to monitor some critical temperature, perhaps the temperature of some sensitive chemical, you will probably need the more accurate calibration made possible by ten-turn trimpots.

Even with the best ten-turn trimpots you can find, don't expect miracles. This simple circuit is rather crude, and would not be at all suitable for very critical applications requiring extremely high precision. The larger the dial face of the meter, the easier it will be to read small differences, but even with a huge meter, you aren't going to get true tenth of a degree accuracy

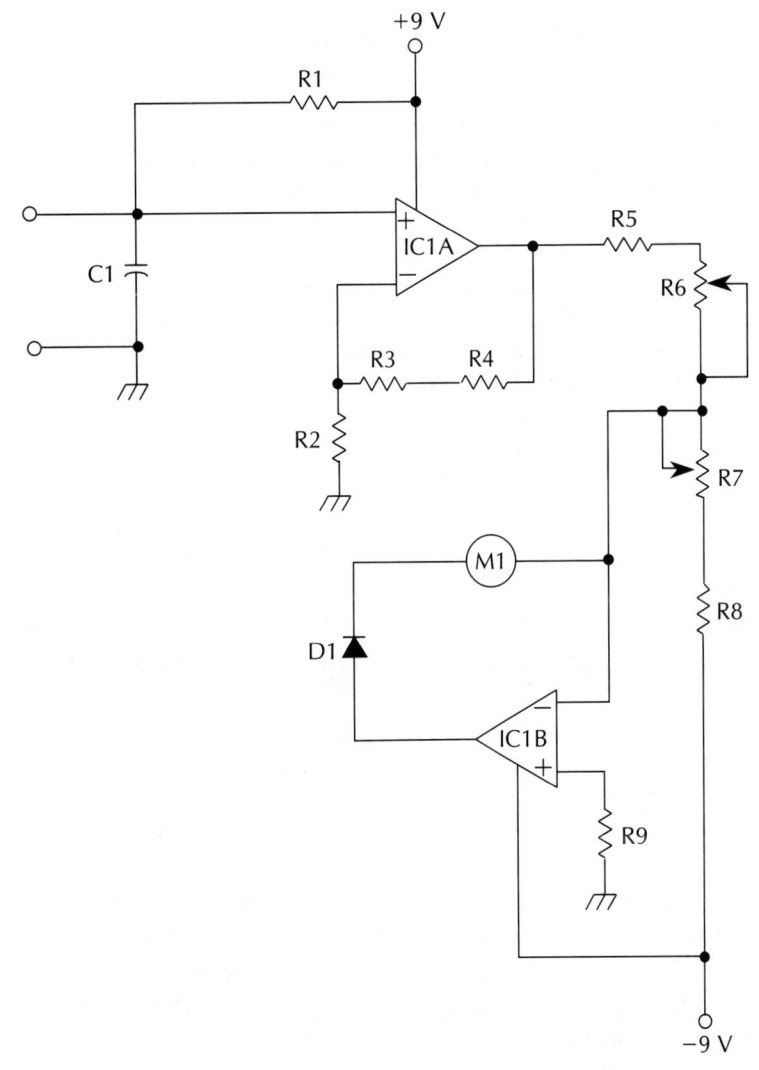

Fig. 2-7 *Project 5. Simple electronic thermometer.*

from this inexpensive project. But it should be more than adequate for most common, general-purpose applications.

In some cases, you might get slightly improved performance by using high-grade precision op amps in this circuit, but the difference probably won't be very great. The various resistor tolerances are probably more important to the overall accuracy of the circuit than the specifications of the op amps. If you've got a couple of high-grade op amps handy, or if they are just a little more expensive than a garden variety op amp, it would probably be worthwhile to go ahead and use the better quality op amp de-

Table 2-5 Suggested parts list for Project 5. Simple electronic thermometer.

IC1	dual op amp (747 or similar)
Q1	NPN transistor (2N3904 or similar) (for probe)
D1	diode (1N4148 or similar)
M1	1 mA meter
C1	0.047 µF (microfarad) capacitor
R1	15 kΩ ¼ W 5% resistor
R2	5.1 kΩ ¼ W 5% resistor
R3, R4	10 kΩ ¼ W 5% resistor
R5	470 Ω ¼ W 5% resistor
R6	500 Ω trimpot
R7	5 kΩ trimpot
R8	2.2 kΩ ¼ W 5% resistor
R9	100 Ω ¼ W 5% resistor

vices. But you'd probably be disappointed if you use a couple of $10 op amp chips in this project, when standard-grade op amps costing less than a dollar will work almost as well.

Because this circuit uses two op amp stages, it makes sense to select a dual op amp chip, such as the 747 or the 1458. This will make the circuit board a little more compact, and will probably lower the overall cost of the project a little. But there is no electronic reason not to use two separate op amp ICs, if you prefer. Just remember to connect BOTH supply voltages to each op amp chip. Notice that this circuit requires both a +9 V and a –9 V supply voltage. Both of these voltages are referenced to circuit ground. Most op amp ICs use a floating ground. That is, there will be no direct connection from the chip to the circuit ground. The +9 V supply voltage is fed to one of the IC power supply pins and –9 V is fed to the other power supply pin on the op amp. If a dual op amp chip is used, you only need to make one set of power supply connections, as shown in the circuit diagram. If separate op amps are used, each individual device must have its own complete set of connections to both supply voltages.

Using the component values suggested in the parts list, the nominal range of this simple electronic thermometer project runs from about –30°F to approximately +120°F. This range should be a more than adequate range for most common general-purpose applications. The exact range of your circuit might vary somewhat, depending on component tolerances, and the characteristics of the specific sensor you use with your project. The temperature sensor for this circuit will be discussed later in this section.

For most practical applications, the temperature sensor will be an external probe. It is not mounted on the circuit board itself, which is why it is not shown in the schematic diagram of Fig. 2-7. Only the connection points are illustrated here. For the time being, you will ignore the temperature sensor itself, except to note that it functions electrically like a variable resistance, which varies in direct proportion to the sensed temperature. If you show the sensor as a variable resistor, as shown in Fig. 2-8, you can see it forms the lower half of a simple resistive voltage-divider network, with resistor R1. Because the value of resistor R1 is constant, but the resistance of the sensor (Rx) varies with the temperature, the voltage seen at the noninverting input of IC1A will be directly proportional to sensed temperature. Capacitor C1 functions as a simple filter, to smooth out any sharp noise spikes in the input signal. The exact value of this capacitor is probably not too critical for any practical applications.

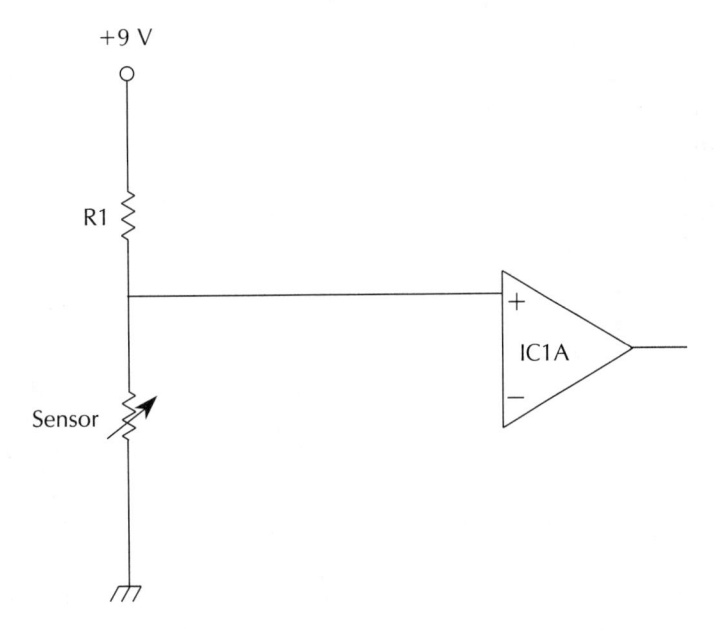

Fig. 2-8 *Conceptually, the sensor in the circuit of Fig. 2-7 acts like a variable resistance in a voltage-divider network.*

Op amp IC1A is wired as a simple noninverting amplifier circuit, with a gain of about 5.

Trimpots R6 and R7 interact to calibrate the electronic thermometer circuit. Notice that these two trimpots are part of another simple resistive voltage divider network between the output of IC1A and the −9 V supply voltage. Resistors R5 and

R8 are also part of this voltage-divider network. The voltage at the junction of trimpots R6 and R7 is tapped off to act as the input to op amp IC1B.

Op amp IC2 is used in its inverting amplifier mode, as a simple current meter circuit. A 0 to 1 mA dc milliammeter (M1) is placed in the feedback loop of the op amp, in place of the usual resistor. Diode D1 protects the meter from damage due to an incorrect polarity resulting from an out of range temperature measurement.

Now you are ready to calibrate the electronic thermometer project. If the sensor is held at a temperature of −30°F, the current through R5 and R6 should be exactly equal to the current flowing through R7 and R8. Trimpots R6 and R7 should be adjusted for this condition. To calibrate the circuit, refrigerate the sensor to −30°F. The more exact this temperature is, the better the calibration of your project will be. You can double check the calibration temperature with some other (previously calibrated) temperature-measurement device.

Give the sensor a little time (at least ten to fifteen minutes) to adjust to this cold temperature. Carefully adjust trimpot R6 for a value of 390 Ω, with the aid of an accurate ohmmeter. You could use a fixed resistor of 390 Ω here, but this does not allow for normal component tolerances. Measure the exact value of resistor R5. Add or subtract any variation from the nominal 470 Ω value to the 390 Ω target value of trimpot R6. For example, if the measured value of resistor R5 is 457 Ω, then it is 13 Ω too low. Add this 13 Ω to the target value for trimpot R6, which should be set for a value of 403 Ω. Similarly, if the measured value of resistor R5 is a little high, say 488 Ω, the extra tolerance resistance should be subtracted from the target value of trimpot R6 (372 Ω).

In a noncritical application, you could substitute a fixed 390 Ω resistor for trimpot R6, but expect some (minor) loss in overall accuracy of your electronic thermometer project.

The next step in the calibration procedure is to adjust trimpot R7 for a reading of zero on the current meter (M1). You might find you have to readjust the value of R6 slightly.

If your intended application isn't too critical, this should be sufficient calibration. But to make sure, you can now heat the sensor up to exactly 120°F. Again, wait a few minutes to give the sensor time to stabilize to the new temperature. Now you should get a reading of exactly 1 mA (full scale). You might have to readjust trimpot R6 slightly to accomplish this. For perfect calibration, you will need to go back and forth between the

−30°F and +120°F measurements several times. This will take some time and patience, but it is not particularly difficult.

For most applications, it will be helpful to replace the dial plate of the milliammeter (M1) with a faceplate calibrated in degrees Fahrenheit, rather than current (milliamperes). This is not difficult to do. The old 0 mA point is marked −30°F, and the old 1.0 mA position is labelled +120°F. Fortunately, the circuit response within its range is essentially linear. Divide the rest of the dial plate into convenient equal units, and add the appropriate labels. For example, if you divide the meter range into 15 equal units, each unit will indicate a difference of 10°F.

To double check the calibration, you can check the circuit at some intermediate known temperature(s), and make sure you get the correct reading on the modified meter.

If your application is not too critical, you can cheat a little on the calibration procedure. It is a little difficult to get a solid −30°F (or +120°F) control temperature. But it is very easy to get a control temperature of +32°F, which is the freezing point of water. Water and ice can coexist simultaneously in a single container at this temperature. Place the sensor in an ice-point bath. An *ice-point bath* is simply a glass beaker of cold water and ice in roughly equal quantities. Wait a few minutes to give the temperature time to stabilize. Set trimpot R6 for a value of 390 Ω, plus or minus any compensation for the tolerance in the value of resistor R5. Then leave this trimpot alone. Now, you can adjust the calibration trimpot (R7) for a reading of 32°F on the meter. Because of the linear nature of this electronic thermometer circuit, if the reading is correct for one temperature within the acceptable range, it can be presumed to be at least reasonably correct for all other temperatures within the project range.

This project is designed for use with a very simple temperature sensor. A common, low-power NPN transistor with its base shorted to its collector, so it is a sort of modified diode, as shown in Fig. 2-9. The semiconductor material within the transistor is temperature sensitive. Ordinarily this is a disadvantage, but here it is precisely what you want. Using a transistor in the diode mode (rather than a true diode) maximizes this effect. Essentially, the conductance (the reciprocal of the resistance) of the semiconductor varies linearly with its temperature. The effect is quite small, so the sensor signal is very, very weak. This is why an input amplifier stage (IC1A) is mandatory in this project. If the sensor is mounted off the main circuit board (which would be the case in most practical applications of this project), you

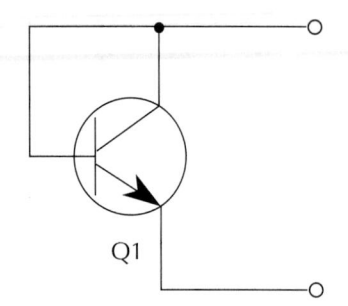

Fig. 2-9 A diode-connected transistor is used as the sensor in Project 5.

should use a twisted pair to connect the sensor to the main circuit board. This will minimize most interference problems. A twisted pair is simply two wires twisted around one another, as shown in Fig. 2-10.

In an electrically noisy environment, the connecting wires should be well shielded.

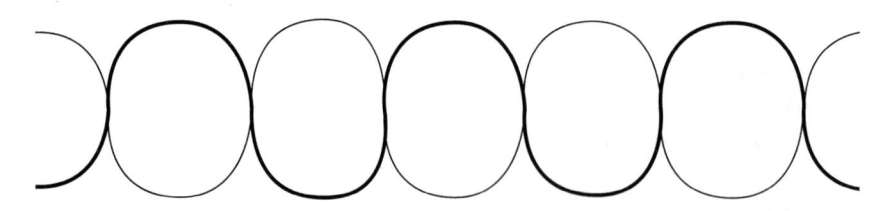

Fig. 2-10 Using a twisted pair to connect the sensor to the main circuit board will help minimize potential interference problems.

Almost any low-power NPN transistor should work as the temperature sensor in place of the 2N3904 transistor recommended in the parts list. However, if you make such a substitution, you should be aware that this might alter the operating range of the project somewhat. For best results, you might want to experiment with different values for resistor R1 to achieve the best possible accuracy over the entire desired range. The 15 kΩ value suggested for this component was selected assuming a 2N3904 transistor will be used for the sensor.

Voltmeter thermometer adaptor

To measure a temperature electronically, you need to convert the detected temperature into some electrical parameter, which has a value that is proportional to the sensed temperature. By converting the detected temperature into a proportional dc voltage, you can use a standard voltmeter section of a VOM or

DMM (volt-ohm-milliammeter or digital multimeter) to perform a lot of the work and to produce the output display. This significantly lowers the overall cost of the project and permits very good accuracy. The voltmeter/thermometer accuracy is largely the function of the accuracy of the voltmeter section. If you use a high-grade VOM or DMM to make the measurements, you can get excellent results. Best of all, assuming you already own a good VOM or DMM, this part of the project (which contains most of the serious circuitry) won't cost you anything at all beyond the minimal cost of a couple of appropriate jacks to plug the temperature to voltage adapter circuit leads into the sockets normally used by the meter test leads.

In practical terms, a temperature-to-voltage converter circuit like this project probably wouldn't be very suitable for a permanent room thermometer or any similar application. It will work just fine in such applications, of course, but you might be less than satisfied on the convenience level. Obviously while your VOM or DMM is being tied up as part of the voltmeter thermometer, you can't use it for any other purposes. But if you buy a second VOM or DMM just for use in this project, that would defeat much of the purpose. Instead of getting high performance at minimal cost, you'll be paying for a number of VOM or DMM functions that you won't even be using.

A project like this is best suited for spot-checking applications. It is particularly good for tracing out potential hot spots in an electronic circuit as it is being serviced. For such applications, the temperature sensor should be mounted in a small hand-held probe that will fit easily into small spaces on a typical circuit board. Be sure the temperature probe is well insulated. In this case, you need not only electrical insulation, as in any test probe, but also good thermal insulation to avoid throwing off the accuracy of the temperature reading by the heat from your own hand as you hold the probe.

The schematic diagram for this temperature to voltage adapter circuit is shown in Fig. 2-11. A suitable parts list for this project is given in Table 2-6.

This project calls for three rather unusual semiconductor components. Make sure you have a source for all the required components before you start buying parts for this (or any other project), or you might end up very frustrated and unable to complete your project. The electronics industry is a rapidly changing field, and a device that was commonly available last month might be obsolete and impossible to locate next month.

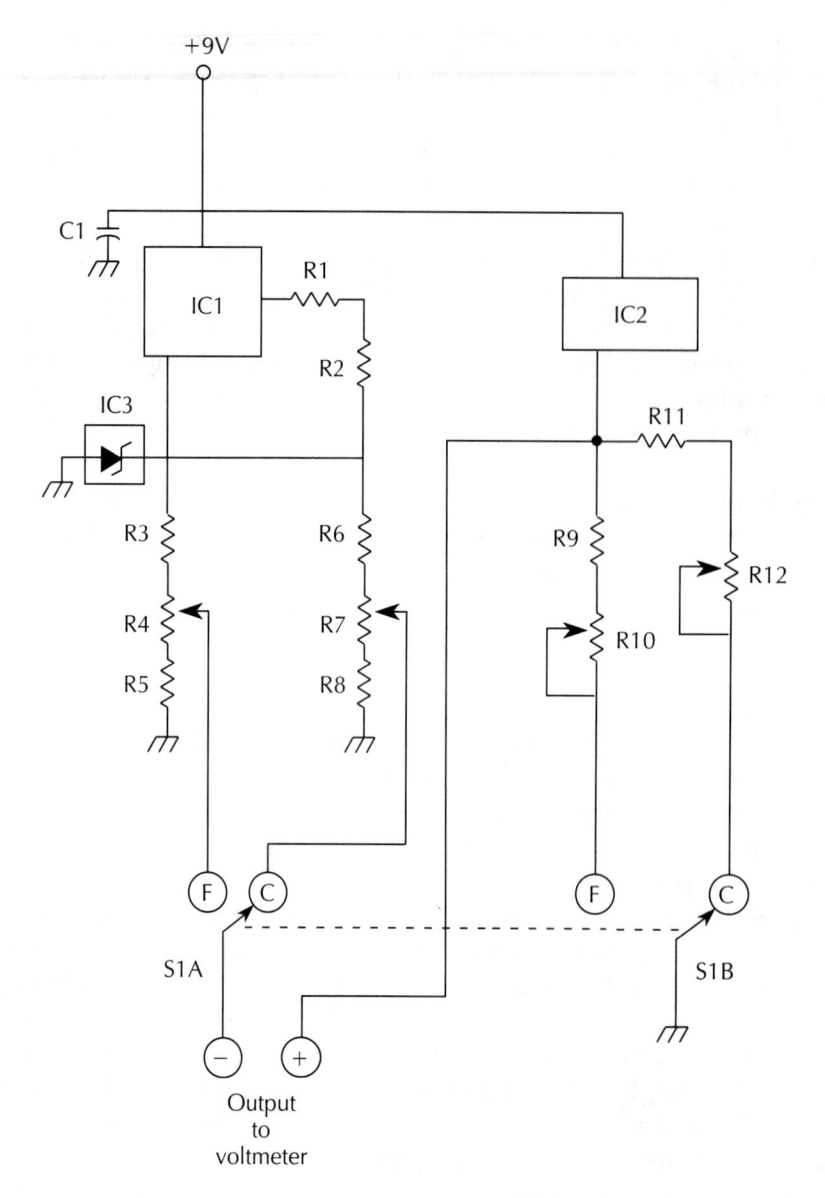

Fig. 2-11 *Project 6. Voltmeter thermometer adaptor.*

Unfortunately, there is not much anyone can do about this type of problem. The components mentioned in this book are normally readily available components, but there are no guarantees about what will be available by the time you try to build the project.

Advanced electronics hobbyists might be able to make suitable substitutions for one or more of these critical components, but such substitutions should not be attempted unless you

Table 2-6 Suggested parts list for Project 6. Voltmeter thermometer adaptor.

IC1	LM342 current source
IC2	AD590 temperature sensor
IC3	LM329DZ voltage reference
S1	DPDT switch
C1	0.05 µF capacitor
R1	22 Ω ¼ W 5% resistor
R2	8.2 Ω ¼ W 5% resistor
R3	10 kΩ ¼ W 5% resistor
R4, R7, R10, R12	5 kΩ trimpot
R5, R8	6.8 kΩ ¼ W 5% resistor
R6	3.9 kΩ ¼ W 5% resistor
R9	15 kΩ ¼ W 5% resistor
R11	8.2 kΩ ¼ W 5% resistor

know what you are doing. Mistakes could quickly fry some expensive ICs, and possibly even damage your VOM of DMM if it is connected to the incorrect circuit's output. Even without any substitutions, it is a good idea to carefully check and double check all connections throughout the circuit before applying power and trying it out.

IC1 is an LM334Z current source. A discrete constant current source built around a transistor should work as well.

IC2 is a specialized device designed specifically for temperature measurement applications like this one. It is the AD590 temperature sensor. Other temperature sensors should work okay, but you might have to make some significant changes in the passive component values throughout the circuit to compensate for operational differences in an alternate temperature sensor.

IC3 is an LM329DZ voltage reference. This device looks much like a common zener diode, but it includes some internal circuitry to operate more precisely. If you can't find this component (or an equivalent from another manufacturer), you could substitute a standard 6.9 V zener diode, although there might be some minor loss of accuracy in the measurements taken by the circuit. Still, the overall accuracy should still be quite good, and if you can't find the LM329DZ, a circuit built around a standard zener diode is certainly better than no circuit at all.

Nothing else in this circuit should be particularly difficult to find, although resistors R1 and R2 might take a little more effort to locate than most other resistors, because their values are so small. In a pinch, you could make up the required resistance(s) with parallel-series combinations of more readily available resistors, of course.

The tolerance ratings of the resistors used in this circuit will tend to affect its accuracy. Use resistors with tolerance ratings no more than 5%. High precision resistors might give better accuracy, but they tend to be significantly more expensive. If you have a lot of 5% resistors handy and some free time, you can measure the exact resistance of each component, and select the one that has an actual measured resistance closest to the nominal value specified in the parts list. As far as the circuit is concerned, this will be exactly the same as using a high-precision (low-tolerance) resistor.

Four calibration controls are included in this project—R4, R7, R10, R12. These should all be screwdriver-adjusted trimpots, not front-panel potentiometers. Once they have been calibrated, leave their settings alone. All four trimpots have a maximum value of 5 kΩ. For the most accurate calibration, you should use ten-turn trimpots, even though they will be considerably more expensive than standard single-turn trimpots. The choice is yours, of course. How accurate are the readings called for in your particular application?

This circuit is designed for direct calibration, so the voltmeter will read out the temperature in degrees. You don't have to perform any calculations yourself. Switch S1 permits you to select between degrees Fahrenheit and degrees Celsius. Notice that a DPDT switch must be used for S1. The two switch sections (S1A and S1B) must always be operated in unison, even though they are shown in different parts of the schematic diagram.

Two common forms of temperature measurement are in widespread use, and this project supports both. The two systems of temperature measurement are called the Fahrenheit scale and the Celsius scale. Each is named after its inventor. The Celsius scale is sometimes called the centigrade scale, because there are exactly 100 equal graduations (degrees) between the freezing and boiling points of water. At sea level, the freezing point of water is 32 degrees Fahrenheit or 0 degrees Celsius, and the boiling point of water is 212 degrees Fahrenheit, or 100 degrees Celsius. As you might suspect, one degree Celsius is larger than one degree Fahrenheit. A *degree* is just an arbitrary unit of measurement. To say "the temperature is 54 degrees," is meaningless unless you specify that temperature scale is being used. A temperature of 54°F is rather cool, and 54°C would be almost unbearably hot.

You won't need to perform any manual calculations between the two common temperature scales in working with this project, but you might find it helpful to know how to do so.

You can convert from degrees Celsius to degrees Fahrenheit with this formula:

$$D_f = 32 + \frac{(9 \times D_c)}{5}$$

where D_f stands for degrees Fahrenheit and D_c stands for degrees Celsius. As an example, see what the Fahrenheit equivalent of 54°C is:

$$D_f = 32 + \frac{(9 \times 54)}{5}$$
$$= 32 + \frac{486}{5}$$
$$= 32 + 97.2$$
$$= 129.2°F$$

You can see that this would be almost unbearably hot.

This equation can be algebraically rearranged to solve in the opposite direction. That is, the formula for converting degrees Fahrenheit to degrees Celsius is:

$$D_c = 5 \times \frac{(D_f - 32)}{9}$$

As an example, convert 54°F to the Celsius scale:

$$D_c = \frac{5 \times (54 - 32)}{9}$$
$$= \frac{(5 \times 22)}{9}$$
$$= \frac{110}{9}$$
$$= 12.22°C$$

Incidentally, there is another temperature scale you might find mentioned here and there, especially in scientific literature. This is the Kelvin temperature scale. A change of one degree Kelvin is exactly the same as a change of one degree Celsius. The difference between these two temperature scales is the zero reference point. The Celsius scale, as already mentioned, uses the freezing point of water (at sea level) as its zero reference point. Temperatures warmer than this are positive, and colder temperatures are negative. For example, 17° below 0°C is written as −17°C.

The Kelvin scale, on the other hand, is referenced to absolute zero—the temperature at which all molecular motion stops. Not surprisingly, this is an incredibly cold temperature. Zero degrees Kelvin is equal to −273.2°C or −459.76°F.

There is no such thing as a negative temperature in the Kelvin scale. It is physically impossible for any temperature to be colder than 0° Kelvin, so all Kelvin temperature measurements must be positive.

It is certainly easy enough to convert between the Celsius and Kelvin scales:

$$D_c = D_k - 273.2$$
$$D_k = D_c + 273.2$$

Just for the sake of completeness, you should also be aware of the Rankine temperature scale. It is referenced to absolute zero, like the Kelvin scale, except it uses the same size degrees as the Fahrenheit scale. Zero degrees Rankine is equal to −459.76°F. The Rankine temperature scale is very infrequently used for practical purposes.

When calibrating this circuit, first temporarily break the connection from the V+ supply voltage and IC2. This effectively removes IC2 and resistors R9 through R12 from the circuit—ignore them for now.

To begin the actual calibration, measure the voltage at the junction of IC3, IC1, R2, R3, and R6. You should get a reading of 6.90 V at this point. This voltage should be correct, because the purpose of IC3 is to precisely set this reference voltage. If the voltage at this point is not correct, the circuit will not be able to function properly. If this unlikely event should occur, you either have a defective reference voltage (IC3), or there is a short somewhere in the circuitry.

Assuming the reference voltage is correct, the next step is to set switch S1A to the Fahrenheit position. Measure the voltage from the − test lead output to ground. Adjust trimpot R4 for a voltage of 4.594 V at this point. If you are using an analog voltmeter, you won't be able to set this voltage as precisely as with a digital voltmeter. Also, it will be difficult to set this calibration voltage precisely and accurately with an inexpensive single-turn trimpot. A good-quality ten-turn trimpot is strongly recommended here.

Once R4 has been calibrated, change switch S1A to its Celsius position, and adjust trimpot R7 for a voltage reading of 2.732 V. Of course, the same notes for the last step in the calibration procedure apply here as well.

Now, reconnect the IC2 section of the circuit. Connect the + and − output leads to the test lead inputs of your voltmeter. Set switch S1 to the Fahrenheit position. Warm or cool the temperature sensor (probably in an external probe of some sort) to a known, constant reference temperature. Adjust trimpot R10 so the correct temperature value is displayed on the voltmeter. An ice bath is a convenient reference temperature. Mix equal portions of cold water and ice, and soak the temperature sensor probe in the bath for at least fifteen minutes to stabilize it. With switch S1 set for the Fahrenheit scale, you should get a reading of 32. Adjust trimpot R10 for a correct reading. You can take another convenient reference with water just at the boiling point. This should give you a reading of 212°F, of course. You might have to go back and forth between the two reference temperatures to get fully accurate calibration.

Repeat the calibration procedure with switch S1 set for degrees Celsius. Naturally, this time you should get readings of 0 for the ice bath, and 100 for the boiling water.

If your project shows signs of instability, you might want to experiment with different values for the power supply filter capacitor, C1.

Electronic thermostat

A *thermostat* is a device that acts as a temperature-activated switch. Standard thermostats are electromechanical devices. Usually, a mercury-filled tilt switch is used. It is connected to a coil made up of strips of dissimilar metals, which have different expansion and contraction rates when the temperature changes. This causes the coil to wind tighter, or unwind slightly, moving the mercury switch in either case. When the angle exceeds a certain point, the mercury inside the body of the switch creates an electrical short between the contacts, activating the switch, which controls a heating or cooling unit.

You can achieve pretty much the same function electronically. A simple electronic thermostat circuit is shown in Fig. 2-12. A suitable parts list for this project is given in Table 2-7.

Basically, this circuit is just a simple voltage comparator. The two voltages being compared are derived from a pair of re-

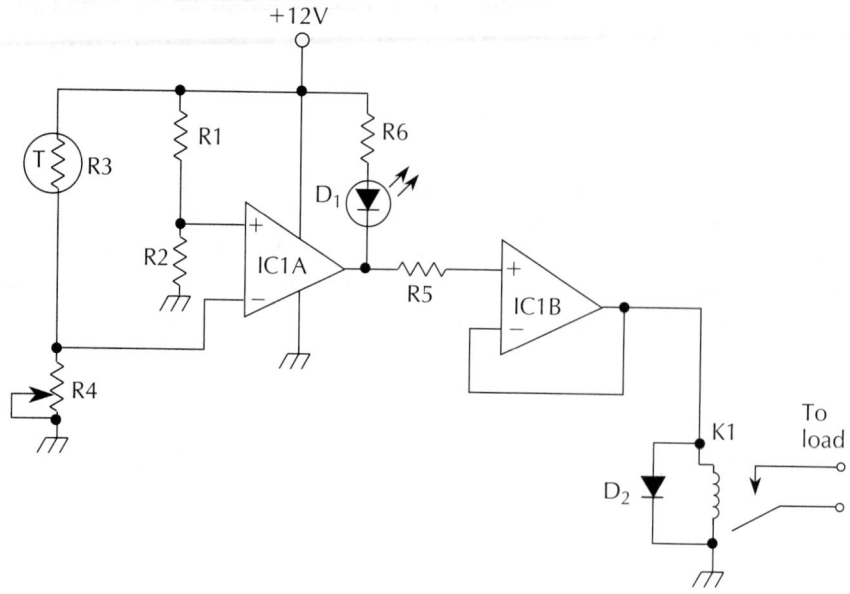

Fig. 2-12 Project 7. Electronic thermostat.

Table 2-12 Suggested parts list for Project 12. Power usage meter.

D1, D2	3.9 V zener diode
M1	VU meter
R1	1 Ω 5 W 5% resistor
R2, R3, R4	10 Ω 1 W 5% resistor
R5, R7	270 Ω ½ W 5% resistor
R6	150 Ω ½ W 5% resistor
R8, R9	0.5 Ω 10 W 5% resistor
F1	4 A fast-blow fuse
F2	15 A fast-blow fuse
S1	SPST switch (rated for 3 A or more)
SO1, SO2	ac electrical sockets

sistive voltage divider networks. The reference voltage is tapped off between two fixed resistors (R1 and R2). The second voltage divider network is made up of a thermistor (R3) and a potentiometer (R4). The thermostat resistance varies with the temperature. The effective sensitivity can be adjusted by the setting of potentiometer R4. If the monitored temperature goes above a specific preset level, producing a higher voltage than the reference voltage at the junction of resistor R1 and R2, the comparator output goes HIGH. The reference voltage is set by the values of R1 and R2, but you can change the effective trip temperature by adjusting sensitivity control potentiometer R4.

When the comparator goes HIGH, the indicator LED (D1) will light up, letting you know that it is working. If you don't need a visual indication in your particular application, you can omit D1 and current-limiting resistor R6 from the circuit. This omission won't affect anything else in the project.

The output of the comparator is fed through a noninverting voltage follower (IC1B), which serves as a buffer amplifier. When the comparator output is HIGH, the relay will be activated, and the heating device connected to the relay normally open contacts will be turned on. If the monitored temperature is below the critical reference point, the comparator output will be LOW, the relay will be deactivated, and the heating device will be turned off.

You are assuming here that the thermostat is controlling some sort of heating device. If you want to use this circuit to control a cooling device, it is just a simple matter of using the relay normally closed contacts. When the relay is deactivated (by a sensed temperature above the reference point), the cooling unit will receive power. When the sensed temperature drops below the reference point, the relay will be activated, opening up the normally closed contacts and disabling the cooling unit.

You might even want to connect a heating device to the normally open contacts and a cooling device to the normally closed contacts, although this isn't really recommended for this simple circuit. One of these two devices will be on at all times. There is no hysteresis in this simple electronic thermostat circuit. If the temperature is close to the critical reference level, there might be a tendency for it to oscillate back and forth between the heating and cooling units.

A different way to modify this electronic thermostat project for use with a cooling unit, instead of a heating device, would be to reverse the positions of the thermistor (R3) and the sensitivity control potentiometer (R4). This will cause the comparator to respond to temperature changes in exactly the opposite way as in the original version shown in the schematic diagram.

Diode D2 protects the delicate relay coil from possible damage due to back EMF (electomotive force) during switching.

In using this project, remember that most practical heating and cooling devices draw fairly hefty currents. Even with the buffer amplifier stage (IC1B), the circuit might not be able to control a large enough relay directly. In such a case, two relays can be cascaded, as shown in Fig. 2-13 so a smaller relay can control a larger relay. Although the intermediate voltage source is shown here as a battery, any suitable voltage source might be used, even the supply voltage for either the control circuit or for the load might be used, depending on the requirements of the secondary relay coil.

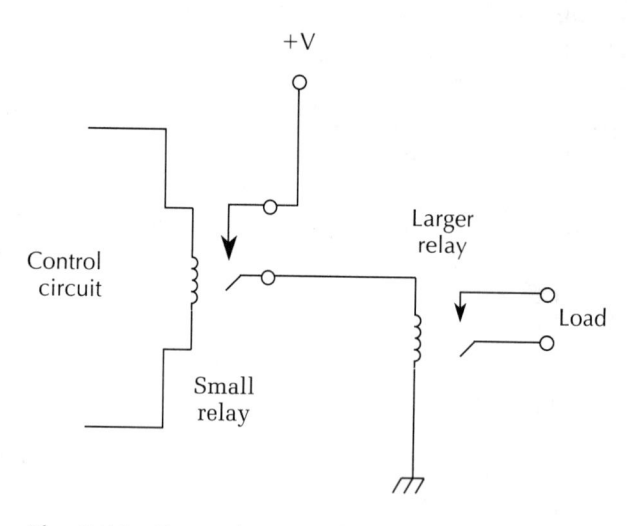

Fig. 2-13 *Two relays can be cascaded to drive a larger load.*

The parts list suggests using half of an LM324 quad op amp IC for this project. This leaves two op amp sections unused. They might be used in some other circuitry as part of a larger system, or they can simply be left unused. Of course, you could use a dual op amp chip, or two separate single op amp ICs. Unfortunately, most (but not all) of these devices require a dual polarity power supply, but the LM324 can be operated from a single-ended power supply. The LM324 is inexpensive and widely available. There is no need to use expensive high grade, low-noise op amps in this circuit. You won't notice any practical improvement in the functioning of the project.

Even though IC1A is being used as a comparator, it is not recommended that you use a LM339 quad comparator, or any other dedicated comparator device. It will work fine for IC1A, of course, but a dedicated comparator would not be suitable for IC1B. Because you must use a general-purpose op amp for IC1B, you might as well use the same type of device for IC1A too.

This circuit is very simple, but quite effective. It does a good job as an electronic thermostat at a very low price.

Temperature equalizer

A standard thermostat is a closed-loop type of automatic circuit. This simply means there is a continuous, circular path throughout the system. The thermostat monitors the room temperature and controls the furnace or air conditioner. The heat from the furnace (for example) changes the room temperature, which affects the thermostat. The input (room temperature) and output (heater control) are interrelated, affecting each other directly. The concept of the closed-loop system is shown in Fig. 2-14.

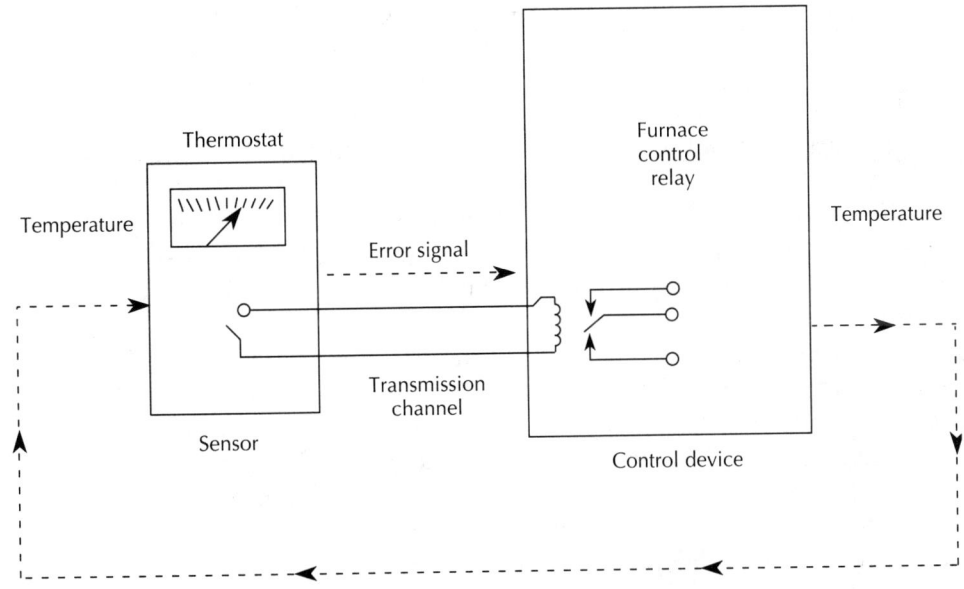

Fig. 2-14   A thermostat is a typical closed loop automation system.

Because of the interrelationship between the input and the output, the closed-loop system can, under some circumstances, exhibit some instability or oscillation. Consider what happens if the thermostat is positioned some distance from the heater duct and there is only limited air circulation in the room. The temperature in the room drops, for whatever reason, so the thermostat tells the furnace to generate more heat. Because of the distance from the heater duct to the thermostat, it will be some finite time before the temperature at the thermostat is high enough for it to shut off the furnace. But most rooms will not have an even temperature across the entire room. Part of the room now might be too hot, even though the thermostat "thinks" the temperature is just right.

When the thermostat finally turns off the furnace, the room as a whole will begin to cool off, especially near any heat leaks, such as doors, windows, or air vents. Assume most of these leaks are also at a significant distance from the thermostat. Again, there will be some finite time delay before the thermostat can sense the drop in the temperature. By this time, parts of the room might be too cold.

You can see how the room's overall temperature oscillates from too hot to too cold and back again, without ever really reaching a happy medium (that is, a truly comfortable, consistent temperature). You not only don't get the proper full benefits of your heating system, you will also tend to waste quite a bit of fuel or power with the unnecessary oscillations.

Of course, an obvious partial solution would be to mount the thermostat close to the heating register, and any significant heat leaks. But this probably won't be possible or practical in most real-world rooms. Even if you could get the thermostat ideally placed, the thermostat can still only monitor its own specific location in the room. Hot or cold spots can still develop in other parts of the room, especially in large rooms.

A better, more consistent solution in most cases would be to increase the motion of the air in the room. Generally speaking, air motion will tend to stabilize the temperature. Warm air will tend to drift towards cold spots, and vice versa. A fan is a simple device for stimulating air motion. In this case, a relatively low-power fan with fairly slowly rotating blades.

The fan will help stabilize the temperature in the room quite inexpensively. But you probably don't want to leave the fan running continuously. This will waste power, and increase ambient noise levels (no practical fan runs truly silently). Moreover, it can cause chilly drafts in some parts of the room, especially directly in front of the fan.

The circuit shown in Fig. 2-15 is designed to control a fan in response to the ambient temperature. The parts list for this project is given in Table 2-8.

Most of the components in this circuit are common, readily available devices and not too critical, so reasonable substitutions are acceptable. The one specialized component here is

Temperature equalizer 71

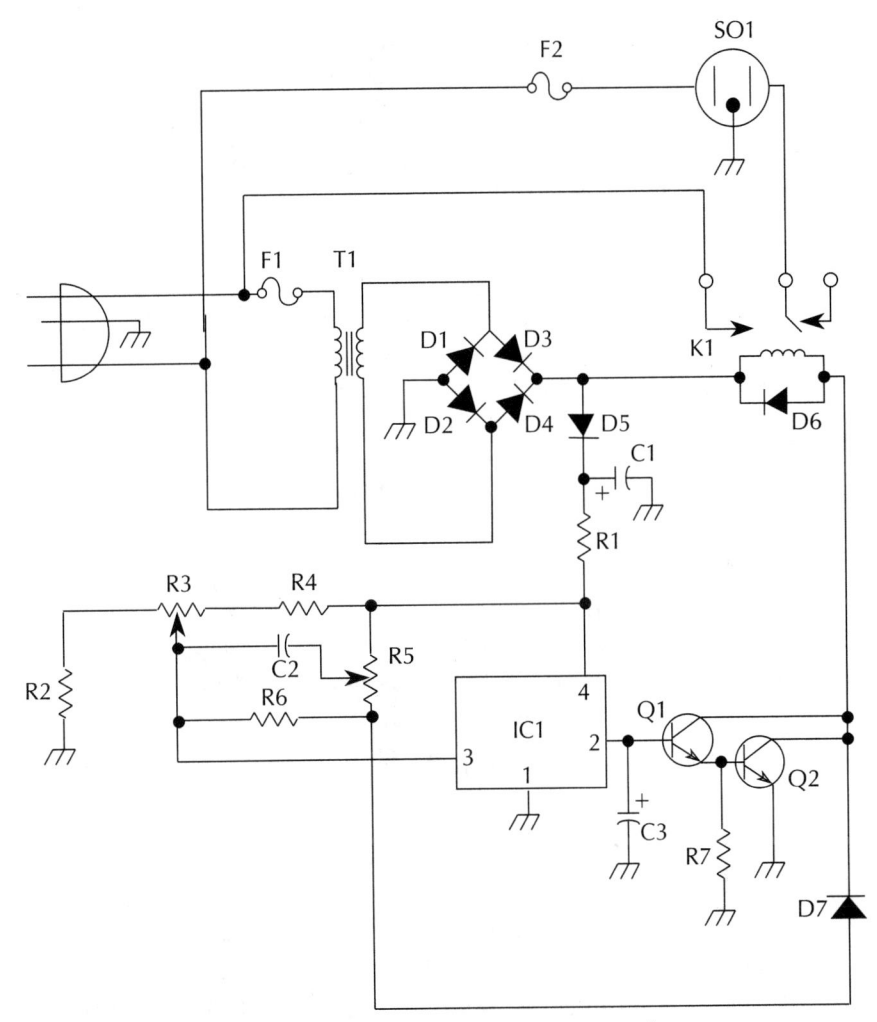

Fig. 2-15 *Project 8. Temperature equalizer.*

IC1, which is used as the temperature sensor for the project. This IC (the LM3911) is designed specifically for temperature-sensing applications. It is made by National Semiconductor and is fairly well available. Unfortunately, as with any specialized electronic component, there is no way to guarantee it won't become obsolete without warning in the future. This can dry up sources very quickly. Make sure you have a source for this part before investing any money into this project, or you might end up frustrated and disappointed.

Transistors Q1 and Q2 are wired as a Darlington pair in this circuit. Their specific operating parameters are not very important in this circuit. Almost any low-power NPN transistors

**Table 2-8 Suggested parts list
for Project 8. Temperature equalizer.**

IC1	LM3911 temperature sensor controller IC
Q1, Q2	NPN transistor (2N3904 or similar)
D1–D6	1N4002 diode (or similar)
D7	1N457 diode (or similar)
K1	24 V relay—contacts to suit load fan
T1	24 V 0.5 A power transformer
F1	½ A fuse
F2	Fuse to suit load (fan)—see text
C1	100 µF 100 V electrolytic capacitor
C2	0.1 µF 35 V capacitor
C3	5 µF 35 V electrolytic capacitor
R1	12 kΩ ¼ W 5% resistor
R2, R4	27 kΩ ¼ W 5% resistor
R3	5 kΩ potentiometer
R5	100 kΩ ¼ W 5% resistor
R6	10 mΩ ¼ W 5% resistor
R7	22 kΩ ¼ W 5% resistor

should work well in this project. They both should be of the same type number.

The relay (K1) should be selected so that its contacts will safely carry the current drawn by the fan. Overrate the relay switch contacts to allow for a margin of safety. If for example, the load will draw a maximum of 1 A, the relay switch contacts should be able to safely carry at least 1.5 to 2 A. Similarly, the output fuse (F2) should be selected specifically for the particularly fan to be used with the system. Do not omit either fuse from the circuit. Do not overrate the output fuse. Typically, this fuse should not be greater than 1 amp or 2 A at the very most.

This circuit carries ac (alternating current) power, and you need to take all relevant precautions. Use adequately heavy wires throughout the circuit. Make sure everything is adequately shielded and insulated. It should be impossible for anyone to touch any circuit conductor (including those nominally at ground potential) while current is flowing through the circuit. Please, don't take any foolish and unnecessary chances.

The temperature that switches on the fan in this project is determined by the setting of potentiometer R3. It's setting is best determined experimentally for maximum comfort. After you have used the project enough to be familiar with it, you might want to make some suitable calibration marks on the front panel of your unit indicating specific positions of this control.

This project is designed to be used together with an existing heating/cooling system using a thermostat. It works equally well whether you are trying to heat or cool the room. The same principles apply either way.

Long-term thermometer

When you are measuring temperature, you often rely on simple spot checking. That is, you look at a thermometer, and see what the temperature is right now. Looking for hot spots or cold spots as in some of the preceding projects in the chapter can help you spot serious flaws in your insulation, but if you really want to know how efficient your insulation is, you will need to monitor the temperature over an extended time. You could check the thermometer periodically, make notes of each temperature reading, and then average them out over the entire test period. What an awful nuisance that would be. The process could be automated by connecting an electronic thermometer to the input port of a computer. The computer could then be programmed to take a reading at certain intervals and perform any necessary averaging calculations. This is better, but it's still rather clumsy and inelegant. For one thing, this would involve tying up a relatively expensive computer for an extended time, without using it at anywhere near its full capacity. Most of the time, the computer would just be sitting there, doing nothing but killing time until the next scheduled reading is due. There's got to be a better way.

Of course there is a better way. (If there wasn't, why would the subject be discussed in a book like this?) The next project is a dedicated circuit designed specifically for this type of application. It is a modified electronic thermometer that gives a continuous read-out of the monitored temperature averaged over time. The circuit can be manually reset at any time to cancel out and forget all previous temperature measurements.

This is a fairly complex project. Rather than attempting to cram the entire schematic diagram onto a single page, it is broken up into four sections, shown in Figs. 2-16, 2-17, 2-18, and 2-19. This makes the circuit details clearer and easier to see. Notice that the subcircuit shown in Fig. 2-18 must be repeated three times in the completed project. There would be little point in illustrating this redundant circuitry in this book. Each of the three duplicate stages are completely identical. In some

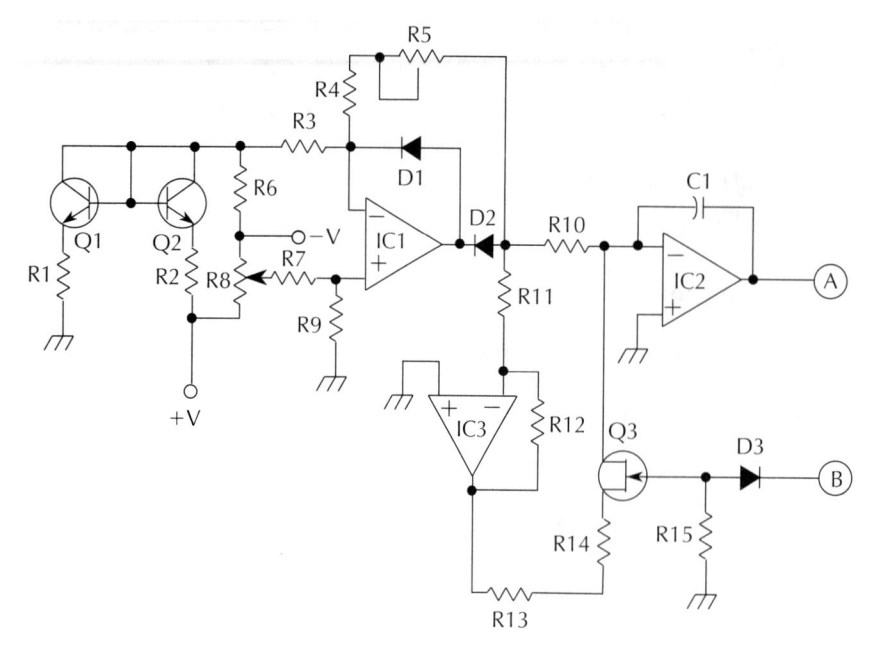

Fig. 2-16 Project 9. Long-term thermometer, part 1.

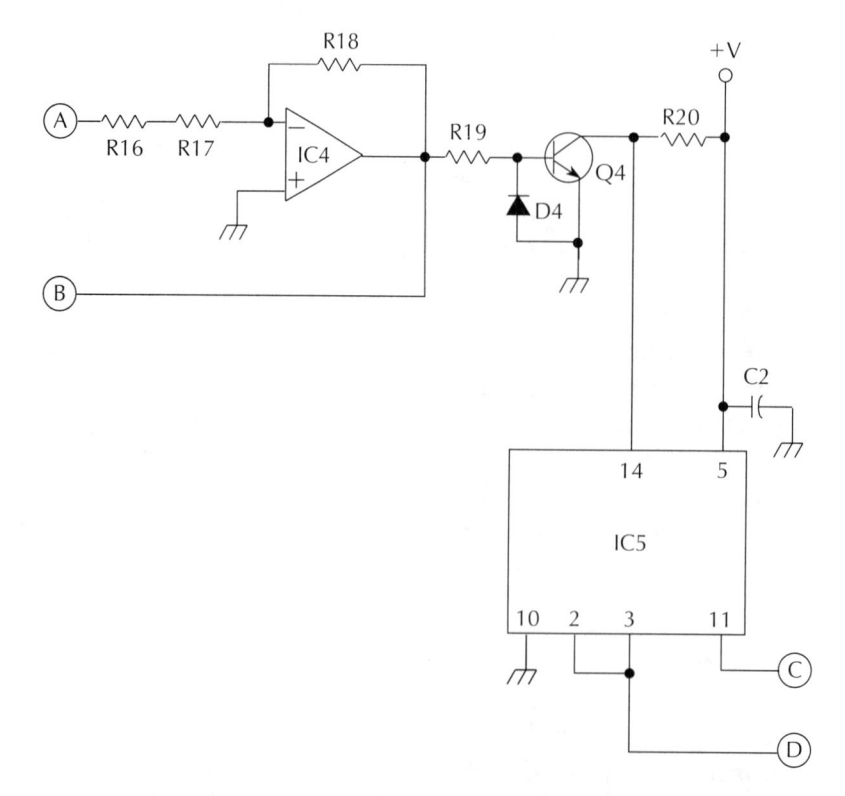

Fig. 2-17 Project 9. Long-term thermometer, part 2.

Fig. 2-18 Project 9. Long-term thermometer, part 3.

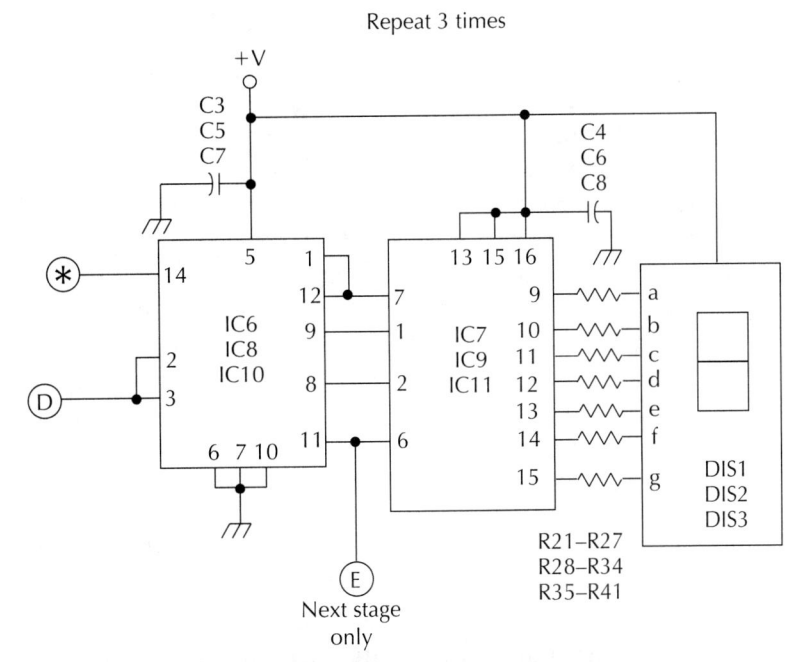

*1st stage (IC6/IC7) connect to C from Fig. 2-16.

2nd stage (IC8/IC9) connect to E from first stage.

3rd stage (IC10/IC11) connect to E from second stage.

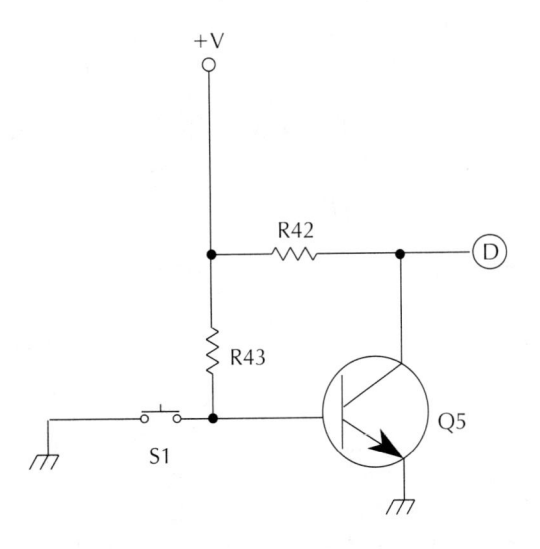

Fig. 2-19 Project 9. Long-term thermometer, part 4.

applications, you might want to eliminate the third of these repeated sections to save a little on the cost of the project.

The averaged temperature is read out on a three-digit display. A decimal point is assumed between the second and third digits, so the displayed value is in the form of xx.x. For example, you might get a read-out value of 74.8 degrees. Many seven segment LED display units have an extra lead for a built-in decimal point. If this is true of the display units you use in your project, it would make good sense to apply a constant voltage to this lead on one of the display units, so the decimal point between the second and third digits is permanently lit. This will make it a little easier to read the averaged temperature values directly from the display.

For many applications, three digits might be a bit of overkill. For most general purposes, there is little, if any practical difference between 82.2° and 82.8°. A rounded reading of 81° or 82° should be close enough for many users of this project. You can omit the third output digit stage for the least significant digit (the one after the decimal point). This will cut the overall cost of the project by a few dollars, because it eliminates a seven-segment LED display unit, two ICs, two capacitors, and seven resistors. Why spend the money if you don't need the extra accuracy of the third digit?

As shown here, the project can measure temperatures up to 99.9°F only. It has no way to display 100.0° or higher. If you intend to use this project in very hot environments (with triple-digit temperatures), you might want to add an extra digit stage for the most significant digit. You probably could get by with a so-called half-digit, which is always either 0 (the display unit is blanked—all segments dark) or 1. Few practical applications for this project will require measurement of temperatures of 200 degrees or more.

All temperature measurements for this project are assumed to be on the Fahrenheit scale.

The complete parts list for this project is given in Table 2-9. Notice that this project does not call for any really unusual components, but because the number of components required is rather large, make sure you have a source for all of the required components before investing any money in the project. Don't set yourself up for unnecessary frustration by finding yourself stymied with a half-finished project that you can't complete because you can't find one critical component.

Table 2-9 Suggested parts list for Project 9. Long-term thermometer.

IC1, IC2, IC3, IC4	op amp
IC5	74C93 binary counter—see text
IC6, IC8, IC10	74C90 BCD counter—see text
IC7, IC9, IC11	74C47 BCD to seven-segment converter—see text
DIS1, DIS2, DIS3	Common-anode seven-segment LED display unit
Q1, Q2, Q4, Q5	NPN transistor (2N3904 or similar)
Q3	FET (Radio Shack RS2028, 2N5457, or similar)
D1–D4	diode (1N4148 or similar)
S1	Normally open SPST push-button switch
C1	10 µF 35 V Mylar capacitor
C2–C8	0.01 µF capacitor
R1, R20–R42	1 kΩ ¼ W 5% resistor
R2, R3, R14	10 kΩ ¼ W 5% resistor
R4	470 kΩ ¼ W 5% resistor
R5	500 kΩ trimpot (scale adjust)
R6, R11, R12, R19	22 kΩ ¼ W 5% resistor
R7	47 kΩ ¼ W 5% resistor
R8	50 kΩ trimpot (zero adjust)
R9	100 Ω ¼ W 5% resistor
R10	270 kΩ ¼ W 5% resistor
R13	120 kΩ ¼ W 5% resistor
R15, R18	100 kΩ ¼ W 5% resistor
R16	12 kΩ ¼ W 5% resistor
R17	39 kΩ ¼ W 5% resistor
R43	4.7 kΩ ¼ W 5% resistor

IC1 through IC4 might be any standard op amps. You can use four separate single op amp ICs, two dual op amp chips, or a quad op amp IC. The quad chip is probably the best choice, because it will make the completed circuit a little more compact, and will probably be a little less expensive than two dual chips or four singles. Of course, each individual IC must have the full power supply connections made to it, or it will not work. Notice that some op amps will work on a single-ended power supply (V+ and ground), but many demand a dual polarity power supply (V+ and V–). When in doubt, check the manufacturer's specification sheet for the particular device(s) you are working with.

The functional requirements for these op amp stages aren't too critical. Almost any standard op amp devices should work fine. Even the lowly 741 will perform well in this circuit. High-grade, low-noise op amps might improve project operation slightly, but the difference will probably be minimal and not really worth the added expense.

IC5 through IC11 must be CMOS type devices. Do not use similarly numbered TTL chips (such as the 7490, or 74LS90). The supply voltages and signal levels in this project would destroy a TTL IC almost instantly.

You might encounter some difficulty in finding the exact type numbers specified in the parts list. But all of these chip functions are pretty standard. You should be able to substitute some other CMOS IC with the same function. You will need to correct the pin numbers, of course. In some cases, some additional minor changes in the circuitry might be required. Such a substitution is definitely not recommended for an inexperienced or casual electronics hobbyist. To make things a little easier for you if you do choose to attempt such a substitution, the pin-out diagrams for the three CMOS ICs are shown in Figs. 2-20, 2-21, and 2-22. (In the circuit, the 74C90 and the 74C47 are each used three times, so there is no need to show redundant pin-out diagrams.)

The seven-segment LED display units must be of the common-anode type. Common-cathode display units will not work in this circuit, as it is shown here. If you are an experienced electronics hobbyist, you should be able to adapt the circuitry for common-cathode display units if you really want to bother with it.

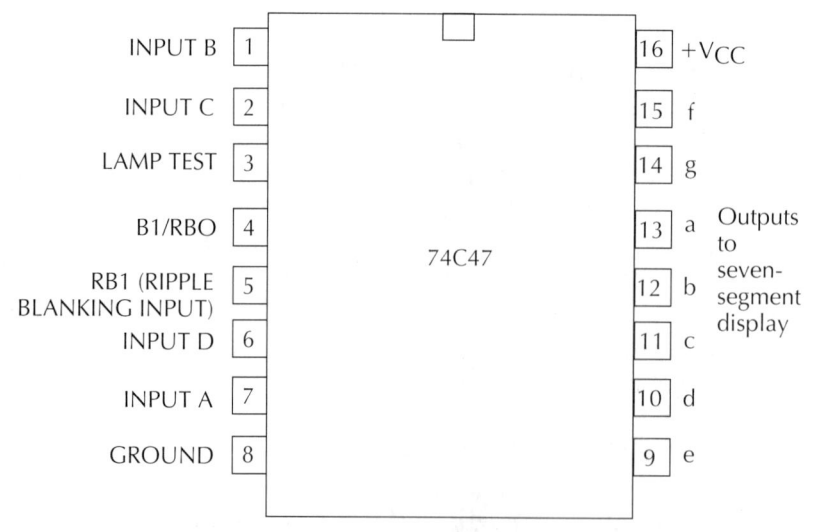

Fig. 2-20 *74C47 BCD to seven-segment converter.*

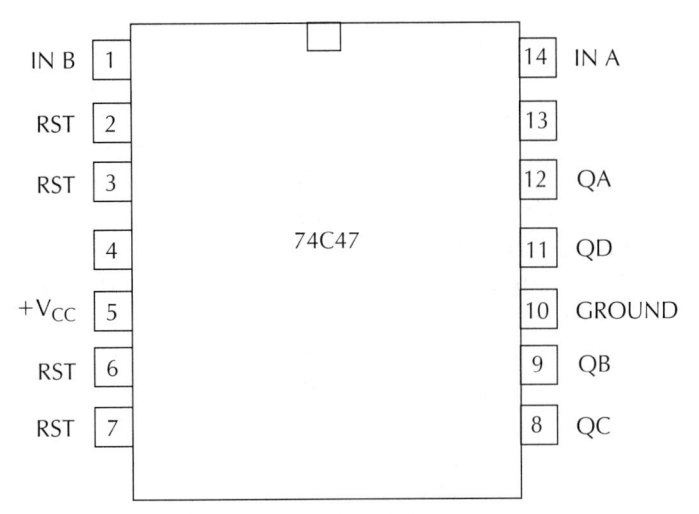

Fig. 2-21 *74C90 BCD counter.*

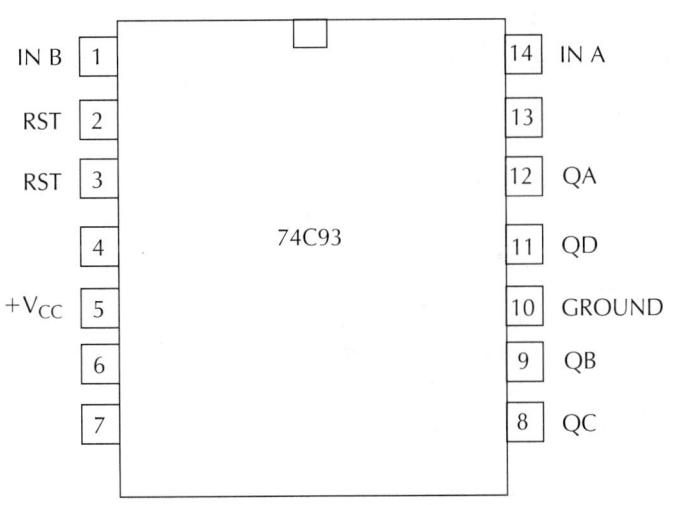

Fig. 2-22 *74C93 binary counter.*

Capacitors C2 through C8 are power-supply filter capacitors to protect the CMOS ICs. They might not be absolutely essential in all cases, but they are cheap insurance, and it is a very good idea to include them in your project. The exact value of these capacitors is not particularly critical, and does not affect the operation of the C2 in any noticeable way. Each filter capacitor should be mounted as physically close to the body of the IC it is protecting as possible. Each IC needs its own individual power supply filter capacitor.

Two NPN transistors (Q1 and Q2) are used to sense the temperature in this project. Almost any low-power NPN transistors should work, although some type numbers might require some changes in a few of the component values throughout the circuit to permit accurate calibration. Both sensor transistors should be of the same type number, or the circuit will be forced out of balance, and it won't be able to give accurate results.

The other transistors in this circuit also aren't too critical, and you should be able to make reasonable substitutions without any problems. Transistors Q4 and Q5 are low-power NPN transistors. They should be of the same type number. They don't necessarily have to be the same type number as the sensor transistors (Q1 and Q2), but using the same transistor type number for all four NPN transistors would certainly make sense and would be convenient. There is no good reason not to use the same transistor type number throughout the circuit, but you could use two different types, if you really want to. The remaining transistor (Q3) is a low-power FET (field-effect transistor).

One of the sensor transistors is mounted outside. It should be exposed to temperature changes, but reasonably shielded from environmental hazards such as rain, snow, hail, and mud. Just use common sense in placing the sensor. The other sensor is mounted indoors. The circuitry compares the difference between the two temperatures, as detected by the sensor transistors, and the difference is continuously averaged over time. Actually, the calculus function known as integration is used to perform the averaging. When the reset button (S1) is briefly closed, the running total is reset back to 00.0, canceling out all previous readings. A new integrated average is started fresh.

Compare the day's average temperature difference with the heating or cooling energy consumed in your home during that period. (The test period does not have to be one 24 hour day, but this is a good choice, at least to begin with.) Check your electrical meter, or keep track of how much fuel (oil, coal, gas, or whatever) you're burning. This will give you a good idea of just how efficient your heating and cooling system is. Are you using more energy than you really need because of poor insulation? Pay particular attention to the energy consumed when there are large average differences between the indoor and outdoor temperatures, as indicated by the project.

Combining this project with one of the heat-leak snooper projects presented earlier in this chapter will help you maximize your energy efficiency, and that could spell considerable savings on your utility bills, not to mention helping the overall ecology of the environment.

In constructing this project, all six subcircuit sections shown in Figs 2-16, 2-17, 2-18, and 2-19 are required for a functional project. Remember, the subcircuit shown in Fig. 2-18 is repeated three times. This repetition is why you only need four schematic diagrams to fully illustrate this six-section project. Connect all points marked A. The same is true for the points marked B and D. This leaves the connection points marked C, E, and *. These inter-stage connections are slightly more complex, because Fig. 2-18 is shown only once, but it should be included three times in the completed project. The connection made to the point marked * depends on which stage you are working. The points marked D are shorted together for all stages. In the first stage, pin 14 of IC6 is connected to the point marked C in Fig. 2-17. In the next stage, you have IC8 in place of IC6. This time pin 14 of IC8 should be connected to the point marked E in the preceding stage. That is, pin 11 of IC6 and pin 6 of IC7 should be connected to pin 14 of IC8. The third and final stage is similar. Pin 14 of IC10 should be connected to point E from the second stage. That is, pin 11 of IC8 and pin 6 of IC9 should be connected to the * at pin 14 of IC10. This is one of those things that sounds terribly complicated when you try to explain it, but is really quite simple in actual practice.

Heat-activated fan controller

The heat-activated fan controller project is ideal for cooling a relatively small area where air conditioning is inappropriate. For example, you might use it in a small work shed that is separated from a main building. A full air-conditioning system would be overkill in such a small space, and it would be ridiculously expensive. Also, if the area you want to cool must be open to the outside air for some reason, an air conditioner would be overworked and wouldn't work very well.

Ordinarily, such an area will be cooled by a fan of some sort. This is fine, but after the fan has been running a while, it might lower the temperature a little too much. You could manually turn the fan on when you feel too warm, and then turn it back off when you feel too cool, but that is hardly an elegant or efficient approach. This is a good application for electronic automation.

The circuit for this project is quite simple, as shown in Fig. 2-23. A suitable parts list for this project is given in Table 2-10.

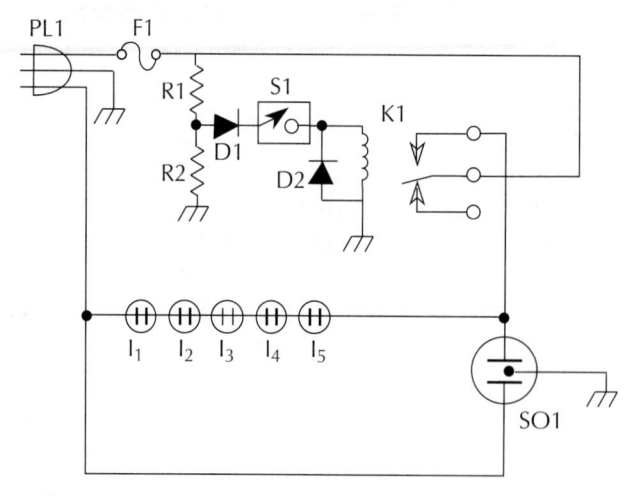

Fig. 2-23 *Project 10. Heat-activated fan controller.*

Table 2-10 Suggested parts list for Project 10. Heat-activated fan controller.

D1, D2	diode (1N4002 or similar)
K1	10 V relay—switch contacts to suit load
S1	thermostat switch—see text
I1–I5	Ne-2 neon lamp—see text
F1	fuse to suit load (2 A maximum)
R1	100 kΩ 1 W 10% resistor
R2	10 kΩ 1 W 10% resistor
PL1	ac plug
SO1	ac socket

The component requirements throughout this project are pretty flexible, and you should be able to make a number of substitutions or modifications to suit your purposes. Almost any standard diodes will work for D1 and D2. D1 serves as a simple half-wave rectifier, and D2 protects the delicate relay coil from potential damage due to back EMF.

The fuse and the relay switch contacts should be selected to match the requirements of the load (fan) driven by the project. A very large fan is not recommended in this application. Anything heftier than a 2 A fuse is not recommended.

Because ac current is used throughout this circuit, make sure your project is completely enclosed and well insulated.

Please do not take any foolish changes that could lead to a fire or a dangerous, even fatal, electrical shock. It's not worth cutting corners.

Not even the values of the two resistors in this circuit (R1 and R2) are particularly critical, as long as R_1 value is about ten times R_2. Be sure that the wattage of the resistors you use will be sufficient. Use Ohm's law and the standard wattage formulas to check this out:

$$I = \frac{E}{R}$$

$$P = EI$$

$$P = \frac{E^2}{R}$$

The only really critical component in this project is the sensor switch S1. It is shown in a box in the schematic diagram to indicate it is not just a standard, manually operated SPST switch. If you use a manual switch, there would be no point in the project at all. It will be exactly the same as manually turning the fan itself on and off, and the added circuitry won't accomplish much of anything. For a useful controller project, S1 must be a thermally activated switch. An old or surplus thermostat using a mercury tilt switch would be ideal. This will permit you to set the desired trip temperature very easily. Use it just like an ordinary thermostat. When the sensed temperature is above the set value, the fan will be turned on. The fan will be turned off when the sensed temperature drops below this preset value.

The neon lamps (I1 through I5) act as noise-suppressing capacitors in this circuit. You can substitute actual capacitors if you prefer. If you can find a good discount package on surplus lamps, the lamps will probably be significantly cheaper than actual capacitors. The neon lamps are just the standard NE-2 type, or almost anything else you might happen to have handy.

If you are using this project to drive a very small fan, you might be able to eliminate one or two of the noise-suppressing lamps shown here. If you are using a larger than normal fan, it might be a good idea to include more neon lamps in the noise-suppressing string to avoid possible premature burn outs due to counter EMF produced by the fan motor.

To use, insert the project plug into any standard ac power outlet and the fan to be controlled is plugged into the project socket (SO1). The built-in power switch in the fan is left in the on position. The fan will be turned on by the control circuit by connecting the input power to the output socket through the switch contacts of relay K1 when S1 senses a too-high temperature. If the sensed temperature is lower than the preset trip value, the relay will be deactivated, breaking the circuit supplying power to the output socket (SO1), effectively turning the fan off.

As in any heating or cooling system, placement of the temperature sensor (the thermostat switch—S1) is important. If the fan blows cool air directly at the sensor switch, the system will shut itself down sooner than if the sensor switch is placed outside the direct air flow from the fan. You will probably want to experiment to determine the best placement for your particular application and preference. Of course, you can move either the sensor switch (S1) or the fan itself, or both—whichever is more convenient.

By using a different type of sensor switch for S1, you can easily adapt this project for many other automation applications. This circuit can control almost any low-power, electrically powered device.

Air-conditioning energy saver

Whenever the subject of energy consumption in the home comes up, the first thing most people think of is "Turn off all lights when not in use." This is fine, as far as it goes, but, perhaps surprisingly, it doesn't really go very far at all. Contrary to popular belief, lighting is not a major drain on energy sources, at least not when compared to other forms of energy use. Turning off the lights faithfully every time you leave the room will shave a few cents off your monthly electric bill, but you're sure not going to get rich from the savings.

The smallest unit on the electrical usage meters used by the power companies is the kilowatt-hour. One kilowatt-hour is the equivalent of using 1000 W of power for one full hour. A 100 W light bulb (the largest size normally used in the home) would have to be on for 10 hours to even show up on the meter. A 60 W bulb would have to be on for over 16½ hours, and a 40 W bulb takes 25 full hours to use up one kilowatt-hour of electricity. According to my current electric bill, I am currently paying about 9½ cents per kilowatt-hour for my electricity. I'd have to leave a 100 W light bulb burning for more than 100 hours to add a dollar onto my monthly electrical bill.

Now, the energy use from lighting is cumulative, and if you

have a large home with many lights, it might be worthwhile to get into the habit of turning them all off when you leave the room. But if you are really interested in energy conservation, this will scarcely be a drop in the bucket.

President Nixon appeared on national television during the big energy crunch of the 1970s to ask citizens to turn off their lights when not in use. A friend and I roughly calculated that if every household in the United States turned off all lights (whether needed or not) for a full 24 hours, the energy saved would not equal the energy consumed to transmit Nixon's speech across the nation. In fairness, most of this energy would have been consumed anyway. The bulk of it was all the television transmitters broadcasting the speech on every (or virtually every) operating station. All of those television transmitters would have been broadcasting some other programming if Nixon had not made the speech. Still, it seems a little ironic to use so much energy advising people to do something that can save relatively minor amounts of energy.

Of course, this was not a formal or scientific study. We had no exact data to work from. We were doing it for fun, and we only worked on it for a couple hours. But the general point is valid. Lighting is far from the biggest energy consumer in society. Compared to the energy consumed by industrial machinery, the power used for lighting is nearly negligible.

In the average American home, the biggest consumption of energy is in the area of heating and cooling the home. Heating or cooling you home for one full day probably uses more electricity than all of the lighting used throughout the home in an entire month. If you live in the south, you know how expensive it can be to run an air conditioner.

Which consumes more energy—heating or cooling a home? That depends on the climate, of course. In some areas you need a lot of heating in the winter, but not much cooling in the summer. In other areas the winters are mild, so not much heating is needed, but the summers are extremely hot and a lot of air conditioning is usually required.

Naturally the specific design of the heating or cooling equipment used will have a considerable effect on the energy consumption involved. As a rough rule of thumb, cooling a given home would tend to use more energy than heating the same area by a comparable amount. This is really a guess—I haven't taken any direct measurements to back up the theory. But there is some good logic to it.

In any system, regardless of its intended purpose, there is going to be some inefficiency and wasted energy. This wasted energy has to go somewhere. In most cases, it will be converted into heat energy and dissipated. In a heating system, this waste heat energy simply helps the system along. In fact, simple electrical heaters rely entirely on such waste heat. Electricity is passed through a resistance element of some sort, which uses up some of the electricity by heating up.

In an air conditioning system, however, the inevitable waste energy in the form of heat not only doesn't help the intended purpose of the system (cooling), it actively works against it. Not only is some of the energy fed into the system wasted as unwanted heat, the system actually has to work a little harder to overcome the effects of its own generated waste heat energy. This fact indicates that, all things being equal, an air conditioning system will probably tend to consume more energy than a comparable heating system.

Now, you've probably guessed after this rather extensive introduction that the next project will be something designed to help conserve energy in an air-conditioning system.

This air-conditioner energy-saver project can be used on almost any central air conditioning system that uses 24 Vdc control circuits. This voltage is more or less standard, but you should be aware there are some exceptions. If your air conditioner uses different control signals, you won't be able to use this project without extensively redesigned circuitry.

For convenience, this project is broken up into two schematic diagrams, which are shown in Figs. 2-24 and 2-25. The complete parts list for this project is given in Table 2-11.

The division of the two parts of the circuit is somewhat arbitrary, and based more on graphics reasons than anything else. But, roughly speaking, Fig. 2-24 shows the circuitry for generating the necessary control signals, and Fig. 2-25 illustrates the switching and interfacing circuitry for connecting this project to your existing air conditioning system via the thermostat. Notice that there are four connection points shown for the thermostat. They are shown in their standard order. Connection point 3 operates the fan and connection point 4 controls the system compressor. Connection point 2 is not used in this project. Make the illustrated connections to the thermostat (after completing and double checking the construction of the project), in addition to any existing connections. Don't disconnect anything that's already part of the system.

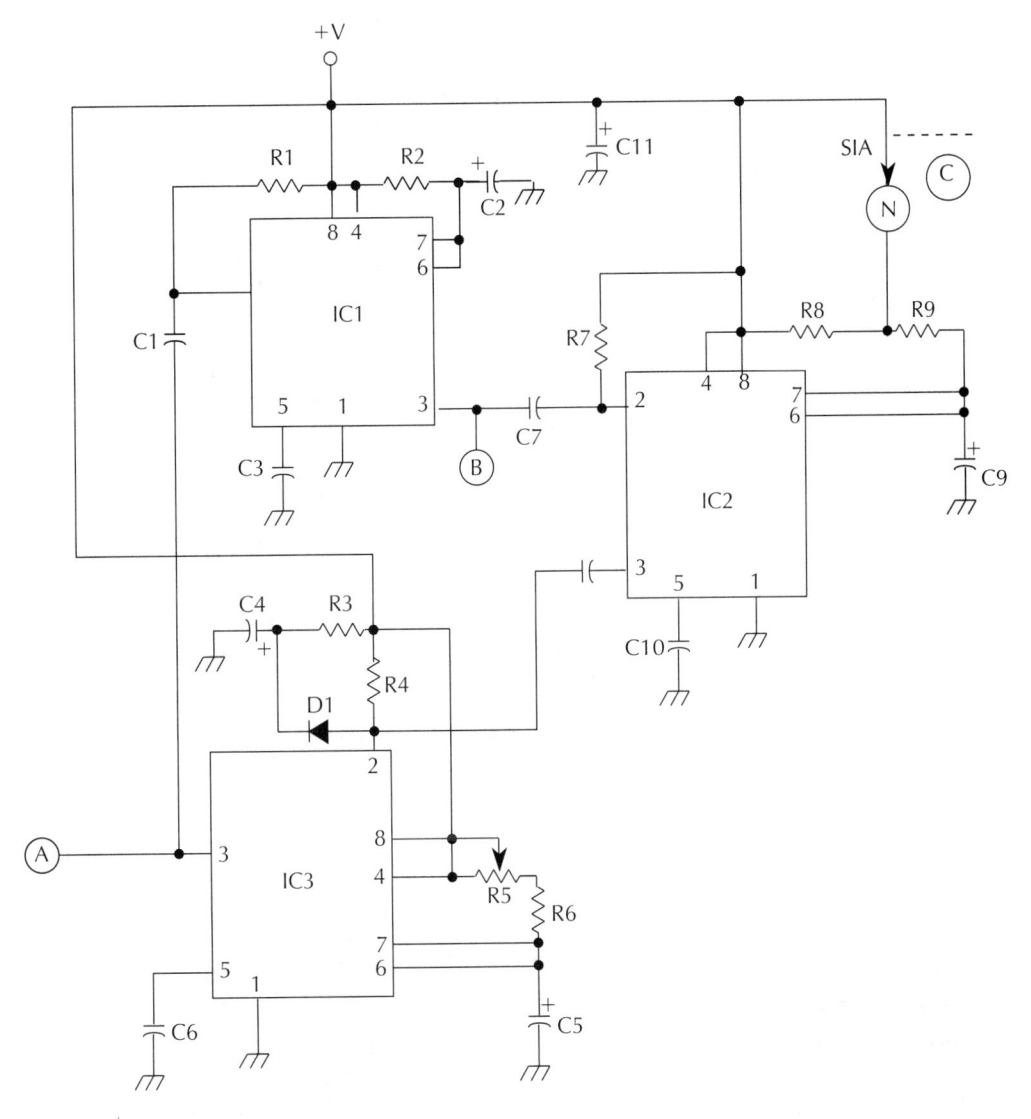

Fig. 2-24 *Project 11. Air-conditioner energy saver, part 1.*

This project uses three 555-type timers to generate its control signals. If possible, use 7555 CMOS timer chips in this circuit, but this choice is not essential. The 7555 is pin for pin compatible with the standard 555, and no changes are required in any of the circuitry to support this substitution.

You might want to combine two of the 555 timers into a 556 dual timer IC. This would not require any changes in the external circuitry, but you will need to correct the relevant pin numbers. For your convenience in making such a substitution, the pin-out

88 Temperature-related projects

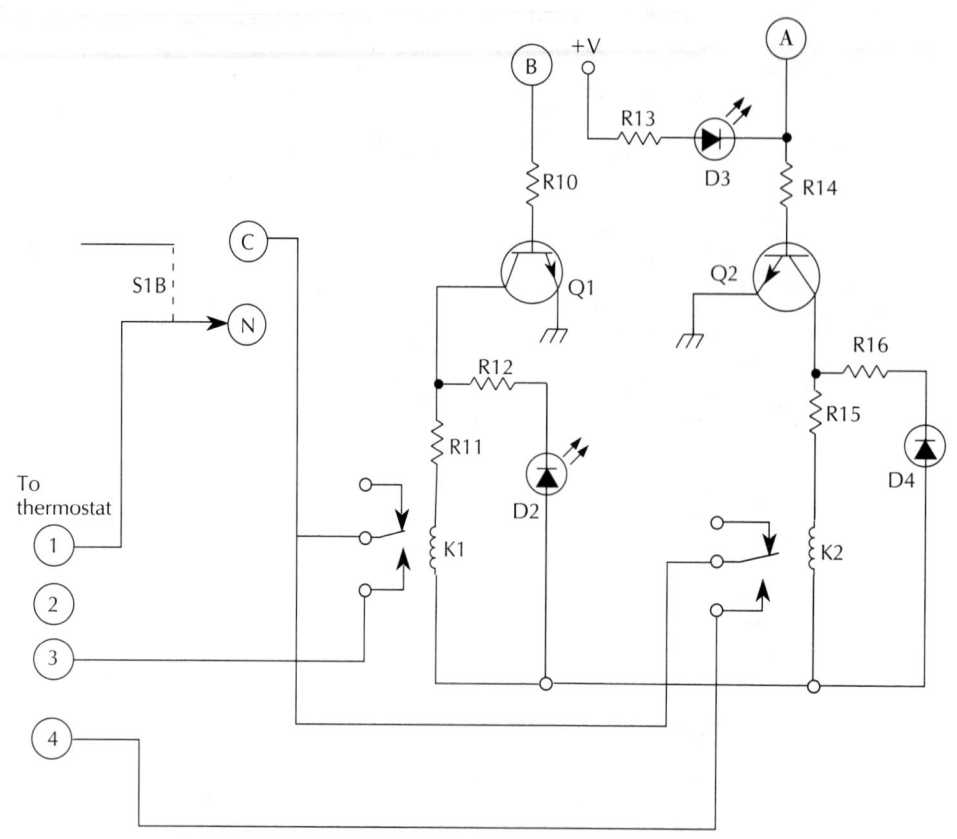

Fig. 2-25 *Project 11. Air-conditioner energy saver, part 2.*

diagram for the 555 (or 7555) timer IC is shown in Fig. 2-26, and Fig. 2-27 shows the pin-out diagram for the 556 dual-timer chip. The 556 is electrically identical to two separate 555s in a single housing (and with common power supply connections).

Almost any low-power NPN transistors can be used for Q1 and Q2. These transistors are used as simple electrical switches, and there are no special requirements for their operating parameters. Some transistors might work better with slightly different values for their base resistors (R10 for Q1, or R14 for Q2). Both transistors should be of the same type number, although even this isn't absolutely essential.

This circuit is designed to operate off a +5 or 6 V supply voltage, but it should work with a power supply voltage up to +9 V or possibly even +12 V. As you can see, this is quite a flexible circuit. Its main function is to operate a pair of relays (K1 and K2), which should be selected to suit the supply voltage,

Table 2-11 Suggested parts list for Project 11. Air-conditioner energy saver.

IC1, IC2, IC3	555 or 7555 timer—see text
Q1, Q2	NPN transistor (2N3904 or similar)
D1	signal diode (1N914 or similar)
D2, D3, D4	LED
K1, K2	Heavy-duty relay (to suit load)
C1, C7, C8	0.001 µF capacitor
C2, C4	10 µF 25 V tantalum capacitor
C3, C6, C10	0.01 µF capacitor
C5, C11	1000 µF 25 V electrolytic capacitor
C9	100 µF 25 V electrolytic capacitor
R1, R3, R4	27 kΩ ½ W 5% resistor
R2, R8, R9	8.2 mΩ ½ W 5% resistor
R5	1 mΩ potentiometer
R6	330 kΩ ½ W 5% resistor
R7	27 kΩ ½ W 5% resistor
R10, R14	4.7 kΩ ½ W 5% resistor
R11, R15	1 kΩ ½ W 5% resistor
R12, R16	3.9 kΩ ½ W 5% resistor
R13	680 Ω ½ W 5% resistor

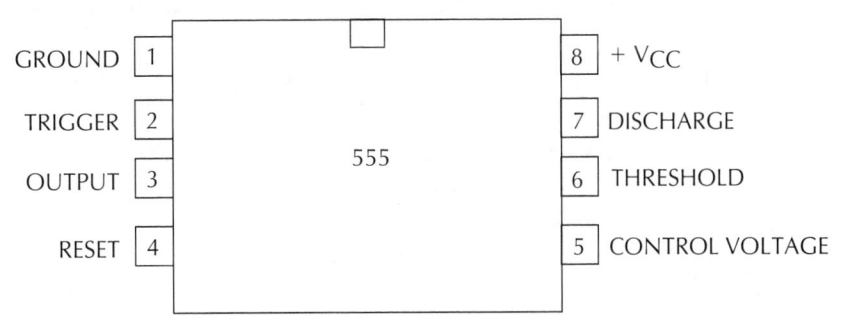

Fig. 2-26 The 555 or 7555 timer IC.

and the load presented by the existing air conditioning system. You can make your relay switching contacts stand up to a heftier load by using DPDT relays instead of the SPDT relays shown in the schematic diagram. Simply wire the two halves of the DPDT switching contacts in parallel, as shown in Fig. 2-28. Each half of this paralleled switch will only have to carry half of the total load current.

Remember, both halves of the circuit (Figs. 2-24 and 2-25) are required for a functional project. Both sub-circuits should

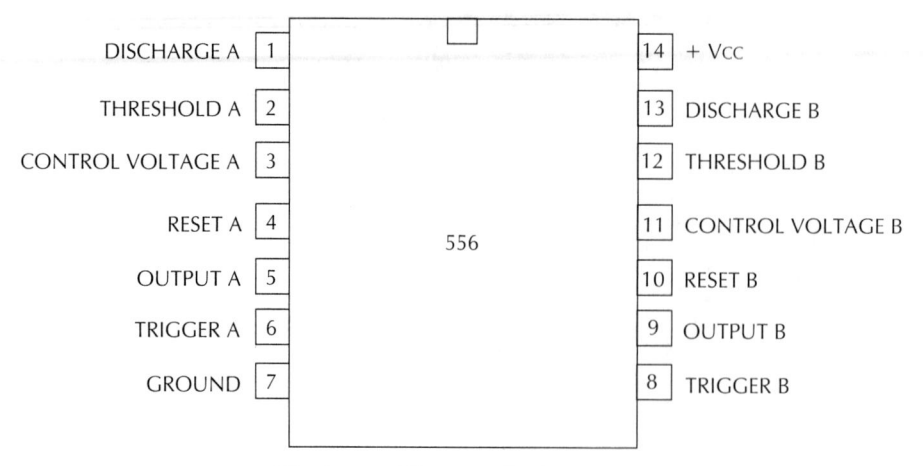

Fig. 2-27 The 556 dual timer IC.

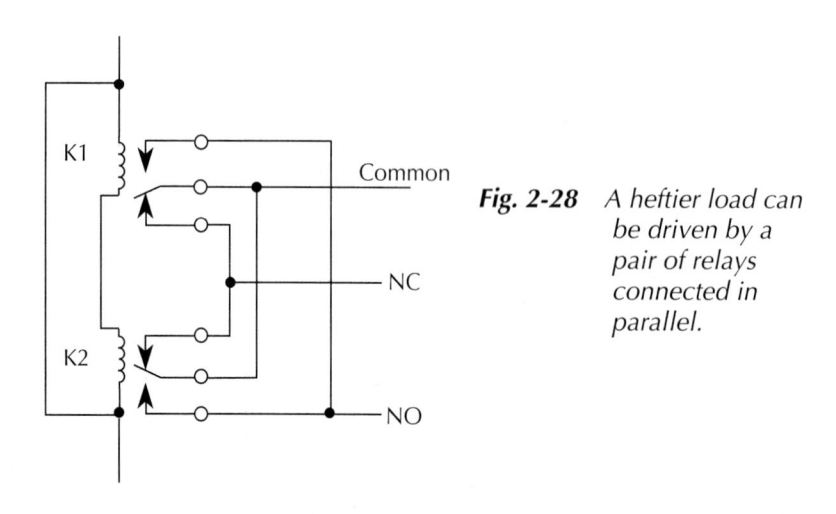

Fig. 2-28 A heftier load can be driven by a pair of relays connected in parallel.

use the same V+ supply voltage, and the points marked *A* and *B* in each diagram should be connected to the corresponding points in the other diagram.

Pay careful attention to switch S1. This is a DPST switch. One of the sections of this switch is shown in Fig. 2-24, and the other switch section appears in Fig. 2-25. Remember, these are two parts of the same switch. The two switch sections are always operated in unison. The two sections ended up in two separate diagrams because that approach made the flow of the circuit easiest to see. When switch S1 is in the position shown in the diagram, the air conditioner functions normally (as indicated by the letter *N* marking this connection point on both switch sections). In effect, the project is disabled when the switch is set to the normal position. The other switch position (marked "C") activates the project and cycles the air conditioner on and off to conserve energy usage.

Without this project (or some similar device), the only way to save on your air conditioning bill is to increase the temperature setting on the thermostat. This will result in a warmer room temperature, but more significantly, it will also tend to result in increased humidity, reduced air movement, and staler air in the room. In other words, it's going to be more uncomfortable in that room with the thermostat set to a higher temperature.

This project is designed to cycle the air conditioner fan and compressor on and off to maximize room comfort, while minimizing the amount of time the air conditioner is actually operating (and consuming energy). Ordinarily, the air conditioner is turned on when the thermostat senses a room temperature higher than its set limit. The air conditioner stays on until the thermostat determines that the room temperature has been lowered sufficiently. This works, but it tends to result in the air conditioner being turned on for unnecessarily long periods, then turned completely off for rather long periods, permitting the humidity and air staleness to increase.

With the air-conditioner energy saver project, the air conditioner is cycled on and off more frequently for shorter periods of time, resulting in more constant humidity and air quality.

If you use this air conditioner energy saver project with a ceiling fan to keep the air moving, even during the air-conditioner off cycles, you will get maximum energy savings. The room can be kept comfortably cool at a higher temperature setting on the air conditioner (resulting in less energy consumption). In other words, by using the ceiling fan to keep the air in the cooled room moving, the air conditioner won't have to work quite as hard for a given level of comfort.

Normally, the air-conditioner fan is operated in unison with the compressor. That is, when the compressor shuts down, the fan is immediately turned off. This project controls the fan and the compressor with separate relays. The fan is kept running for about a minute and a half after the compressor is turned off. This pushes the cool air remaining in the air conditioner's duct out into the room instead of trapping it inside the machinery where it doesn't do you any good. This simple feature offers some considerable extra savings in energy use. The air conditioner had to work (consume energy) to produce that normally wasted cool air, which can now be used, permitting the air conditioner to be set at a lower operating level to produce the same temperature in the cooled area.

Be very, very careful not to get the relay connections mixed up. If you reverse the relays, the compressor will run for a while after the fan has been shut off. This could seriously damage your air conditioner. At best, it will waste all the energy consumed by the compressor during this period, because there is no way for the cooled air to get out into the room—it stays inside the air-conditioner duct.

Potentiometer R1 is used to manually control the main cycling time of the system. For most applications, this potentiometer should be a front-panel control so you can set the project to produce the most comfortable room temperature. For example, you will want to set this control for shorter on cycles when you're not home. There's no reason to cool your house completely when there is no one there to enjoy it. But it is a good idea to keep some continuous cooling going, so it is not unbearably hot and humid when you come home, forcing the air conditioner to work extra hard to bring the environment back into the comfortable range.

You might want to experiment with different cycle times. For the on cycle time, try different values for resistor R6 and capacitor C5. Resistor R9 and capacitor C9 can be changed to alter the off cycle. The fan-delay period is controlled by resistor R1 and capacitor C1. In each case, increasing either or both of the timing components will extend the cycle time for that stage.

Three LEDs (D2 through D4) indicate what is happening in the circuit at any given time. LED D2 lights up whenever the fan is running. LED D4 lights up when the compressor is on, and D3 indicates when the compressor is off. In operation, you should see D2 and D4 lit up and D3 dark while the air conditioner is cycle on. With most air conditioners you will probably be able to hear it running. At the end of the cycle, D4 should go out, and D3 should light up, but D2 will continue to glow for about a minute and a half more. You'll probably be able to hear the air-conditioner fan running, but it will be making much less noise than when the compressor was running. After 90 seconds or so, D2 should also go out, leaving only D3 lit. At some point, the off cycle will time out, and D3 will go dark, D2 and D4 will light up, and the air conditioner will be turned back on for another cycle.

By cycling the air conditioner on and off, you can achieve the same (or even better) comfort level in the cooled room, and using considerably less energy than running the air conditioner continuously in the usual manner.

Power usage meter

If you use an electric heater or air conditioner, you can use this project to keep track of exactly how much power you are using for heating or cooling. Actually, it might be argued that this project doesn't strictly belong in this chapter. There is nothing to limit the use of this project to monitoring the power consumed by electrical heating or cooling units—it can be used to monitor the power consumed by almost any electrically powered device, up to about 1100 W. You won't get accurate readings if the power consumed is less than about 15 to 20 W.

The schematic diagram for this power usage meter is shown in Fig. 2-29. A suitable parts list for this project appears as Table 2-12.

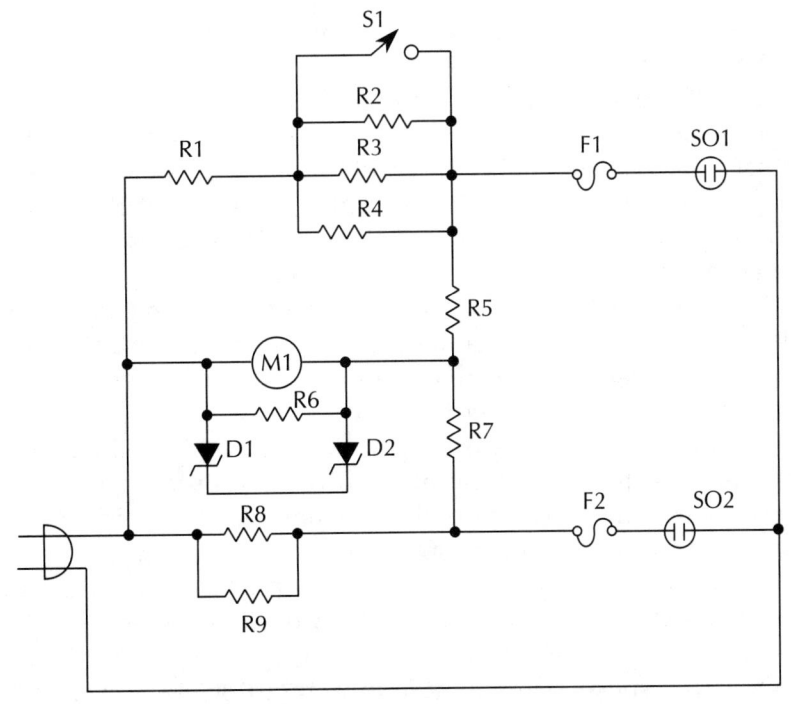

Fig. 2-29 *Project 12. Power usage meter.*

Table 2-12 Suggested parts list for Project 12. Power usage meter.

D1, D2	3.9 V zener diode
M1	VU meter
R1	1 Ω 5 W 5% resistor
R2, R3, R4	10 Ω 1 W 5% resistor
R5, R7	270 Ω ½ W 5% resistor
R6	150 Ω ½ W 5% resistor
R8, R9	0.5 Ω 10 W 5% resistor
F1	4 A fast-blow fuse
F2	15 A fast-blow fuse
S1	SPST switch (rated for 3 A or more)
SO1, SO2	ac electrical sockets

Notice that you are using very low-value resistors in this project. Also notice the wattage ratings are very, very important. The suggested resistor wattages should be considered absolute minimums. You can use a 1 W resistor in place of a ½ W resistor, but do not attempt to make a substitution in the opposite direction. The resistor will burn itself out by the excessive current levels that could be drawn through it, rendering the entire circuit useless. There could also be a serious fire hazard.

You'll undoubtedly notice that resistors R2, R3, and R4 are 10 Ω, 1 W resistors connected in parallel, for a total effective resistance of about 3.3 Ω. So why don't you use a single 3.3 Ω resistor? The answer is in the wattage. The three resistors in parallel share the load of the current flow. The parallel combination can handle 3 W instead of just 1 W. Bypass switch S1 should be rated to handle at least 3 W, too.

Because you are dealing with ac line current in the project, it is vitally important to enclose the entire circuit with plenty of insulation. Please don't take any unnecessary changes with potential fire or electrical shock hazards.

For the same reasons of safety, do not even consider omitting the fuses from this type of circuit. Also, do no ever, under any circumstances, increase the current rating of either fuse. That would defeat the purpose of the fuse altogether.

To use this power usage meter, just plug the electrical device you want to monitor into the appropriate socket (more on this in a moment), and plug the project power cord into a convenient wall socket. Do not attempt to monitor the power usage of more than one device at a time. Only use one of the two sockets in the circuit at a time. Using both sockets at once will confuse the meter read-out totally at best. It could also result in dangerously overloading the circuit.

The meter can be read on three separate ranges. If the monitored device is plugged into socket SO1, and switch S1 is open, the meter range runs from 15 to 60 W. Closing switch S1 ups the range of the meter to 60 to 250 W.

If you want to monitor a larger electrical device, plug it into socket SO2 instead of SO1. In this case, the position of switch S1 is irrelevant. It has no effect on the circuit in this mode. When socket SO2 is used, the meter can read power consumption from 250 to 1100 W.

You can use the original dial plate of the VU (volume unit) meter used as M1, but you will only be able to approximate the power level read-out by the device. For more precise readings, you will want to make up you own calibrated dial plate. This isn't hard to do. Just plug in an electrical device with a consistent, known power rating. Light bulbs are good for the lower ranges. Mark the position of the meter's pointer for a 25 W bulb, a 60 W bulb, a 75 W bulb, and a 100 W bulb. If you can find a floodlamp rated for 250 or 500 W, this will help you calibrate the higher ranges too. Once you've marked a few reference points, you can interpolate intermediate values.

You can also calibrate the circuit using an unknown wattage device, by simultaneously using a voltmeter and an ammeter along with the power usage meter project. Multiply the current value on the ammeter by the voltage value on the voltmeter, and mark the appropriate position on the dial plate of M1.

Remember that most heating and cooling devices do not use a constant level of power. Sometimes they will draw more current, and other times they will draw less. Depending on the device, the power consumption level might be relatively steady, switching between fairly discrete levels, or it might continuously fluctuate over a fairly wide range.

If you are using this project only to monitor the electricity used by heating or cooling devices, you will probably only be interested in using socket SO2. If your intended application will be restricted to such use, you can eliminate socket SO1 from the project. In this case, you can also omit the following components from the circuit: F1, R1, R2, R3, R4, R5, and S1.

These components are used only when the monitored device is plugged into socket SO1. If you make this modification to the circuit, you won't be able to measure power levels below about 250 W.

Similarly, if your application for this project will be restricted to low-power devices only (250 W or less), you can eliminate SO2 from the circuit, along with these components: F2, R7, R8, and R9.

Be careful not to overload socket SO1. If the device plugged into this socket draws more than 250 watts, the meter's pointer will be pinned to the upper end of its scale. It could be bent and permanently damaged by this. No serious harm should be done, because fuse F1 should blow, breaking the circuit before the excessive current can do any damage. A resistor will take some time to burn out from excessive current flow, and there are no delicate semiconductor components in the circuit path in this project.

Similarly, do not overload socket SO2, which is rated for a maximum of 1100 W. Again, the fuse should blow if this limit is exceeded, but the meter pointer might still be pegged and bent before the fuse has a chance to break the circuit path. This will probably only be a real problem if the excessive current is significantly out of range (over about 1500 W or so).

❖3
Liquid-related projects

Ordinarily, water (and other liquids) and electronic circuitry don't mix. At least, they shouldn't. If water (or some other liquid) gets inside an electronic circuit, it can create all kinds of havoc. Most liquids (including water) have a fairly low resistance, so they can easily cause short circuits. And the nature of a liquid is to flow everywhere it can, providing a low-resistance current path between any pair of conductors in the circuit. At best, this will cause the circuit to operate erratically or incorrectly, if at all, until it dries out. More likely, some of the critical (and probably most expensive) components in the circuit will be seriously damaged or destroyed.

In addition, water and many other liquids can cause extensive corrosion of conductors and solder joints. It might also damage the housings of some components. The problems of a paper capacitor, to use just one example, should be obvious enough. So it is vitally important to keep all electronic circuitry dry. So how can there be a chapter on liquid-related projects?

These projects are designed to sense the presence or absence of water and other liquids in varying amounts and for different purposes. Obviously, to accomplish this, you will need to get the sensor wet, but the rest of the circuitry must be kept dry for all the usual reasons. For example, in a flooding alarm project, the sensor should be mounted very low, where flooding is likely to occur, but the rest of the circuit must be mounted high enough that any expected flooding won't reach it. (Of course, there might be an unexpected deluge, but if the flooding is that severe, the damage to your flooding alarm circuit will probably be the least of your worries.) A water-tight housing for any liquid related project is always a good idea.

Commercial water sensors are probably not commercially available, but that is hardly a problem. It is incredibly easy to make your own liquid sensors. All you need is a pair of physically separated conductors. Ordinary dry air has a very high electrical resistance, so there will effectively be an open circuit between the two sensor conductors. But if water, or some other liquid, is simultaneously touching both sensor conductors, there will be a fairly low-resistance current path (through the liquid) between them. The exact resistance will depend on the specific liquid involved and the distance between the two sensor conductors.

It is fairly easy to design a circuit to sense when current is conducted between the two sensor conductors. Several variations on this basic liquid-sensor concept are used in the projects presented in this chapter. A suitable probe is fully described for each individual project.

For the most reliable results, you should select conductor materials for your liquid sensor to be as noncorrosive as possible. Obviously, you don't want a liquid sensor that will rust or otherwise corrode easily if it gets wet.

For safety's sake, please use dc power ONLY for ANY electronic liquid sensing project. Do not use ac line current power. If there is an unexpected short circuit in your project, there could be a very serious shock or fire hazard. The results could literally be deadly. Don't take any foolish chances. Even an ac-to-dc power converter could go bad and send dangerous ac current into the circuitry. Use battery power only, even in a permanent installation. It is much, much safer.

These projects are designed to keep their current consumption requirements as low as possible to make battery power more practical and economical.

Liquid sensor

In some applications it might be useful to have some automated means of determining when the liquid level in a tank (or other container) has been filled past a specific point. Automation systems would be an obvious area of application for such a liquid-sensing device.

A project like this could also be very useful for visually impaired people. It can warn them when pouring that the glass (or whatever) is full. Otherwise, they would have to use a finger to feel when the poured liquid has reached the desired depth. Obviously, using one's fingers is not a very desirable method when pouring hot coffee or a similar liquid. Also, it might be considered a bit rude for a host or hostess to have a finger in drink being served to guests. An electronic liquid probe, such as this project, provides a much better solution.

The schematic diagram for the liquid-sensing alarm circuit is shown in Fig. 3-1. The suggested parts list for this project appears in Table 3-1.

As you can see, this is a very simple project. Most of the work is done by IC1, which is an LM3909 LED flasher/oscillator chip. This device is quite widely available, and it should continue to be available for some time to come. It is a very simple device, with a surprisingly wide number of applications. In this

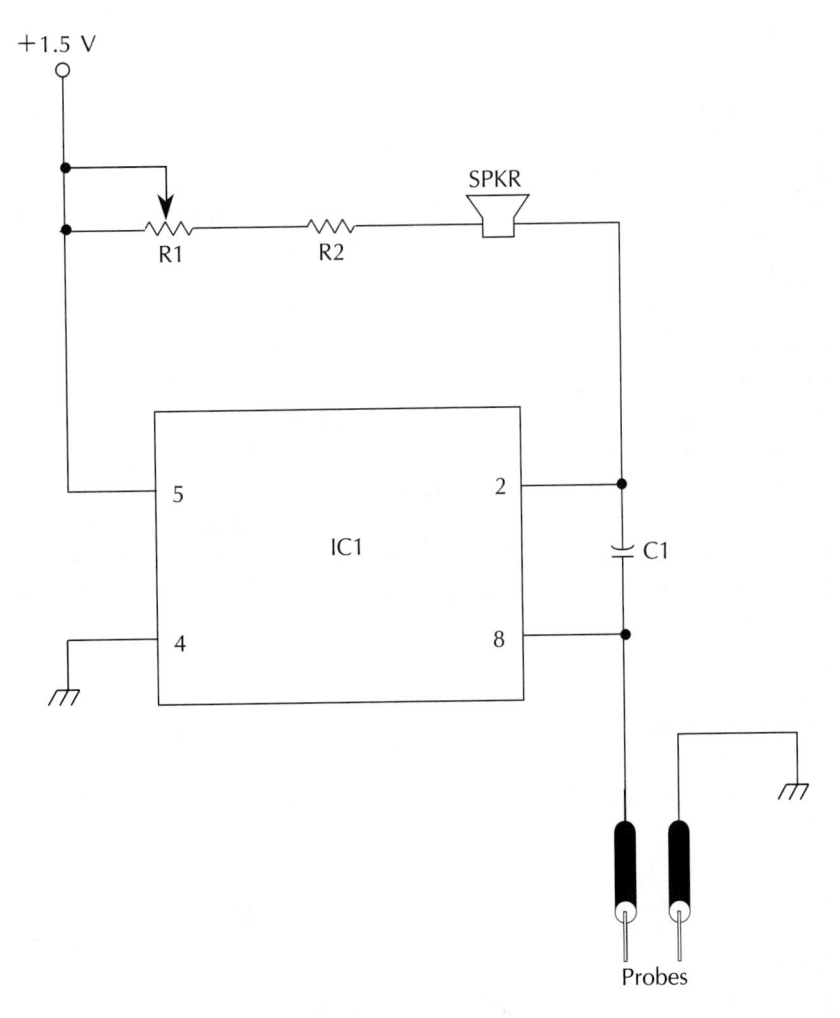

Fig. 3-1 *Project 13. Liquid sensor.*

Table 3-1 Suggested parts list for Project 13. Liquid sensor.

IC1	LM3909 LED flasher/oscillator
SPKR	small 8 Ω speaker
C1	0.01 µF capacitor
R1	100 kΩ potentiometer
R2	33 kΩ ¼ W 5% resistor

project the LM3909 is used in its high-frequency mode, in which it acts like a simple audio oscillator.

One particular advantage of the LM3909 LED flasher/oscillator IC is its exceptionally low power-supply requirements. This circuit can be operated from only 1.5 V. A single AA or AAA battery cell can be used to power your liquid sensor project. The current drain consumed by the LM3909 is virtually negligible. It is little more than the normal leakage current of a battery sitting unused on a shelf. In other words, the battery life in a typical LM3909 circuit under continuous operation would be almost as long as the ordinary shelf life of the battery.

None of the few external component values in this circuit are particularly critical. The basic frequency of the tone is determined primarily by the value of capacitor C1, and the inherent characteristics of the specific type of liquid being monitored. You might want to try experimenting with alternate values for this capacitor to achieve the most pleasing effect.

Potentiometer R1 acts as a volume control for the alarm. Resistor R2 is included in the circuit to prevent the volume control from being set to 0, possibly overloading and damaging the speaker. If a manual volume control is not needed in your application, you can replace R1 and R2 with a single small-value fixed resistor.

In some applications (such as a pouring aid for the visually impaired, as described above), a transistor radio earphone might be a better choice for the output device than a loudspeaker. That way, only the user hears the tone.

Basically, this circuit is just a simple audio oscillator. Most liquids (especially water) have a relatively low electrical resistance. When the probes are dry, there is a very high resistance between the two probes. In effect, without a liquid to form a current path between them, there is essentially an open circuit between pin 8 of the LM3909 and ground. But when an electrically conductive liquid touches both probes, it electrically functions like a resistor connected between the probes (and therefore, pin 8 and ground), as shown in Fig. 3-2. This forces the frequency of the oscillator to rise, so the pitch of the tone heard from the speaker (or earphone) suddenly goes from low to high. With most liquids, the difference in pitch will be quite unmistakable.

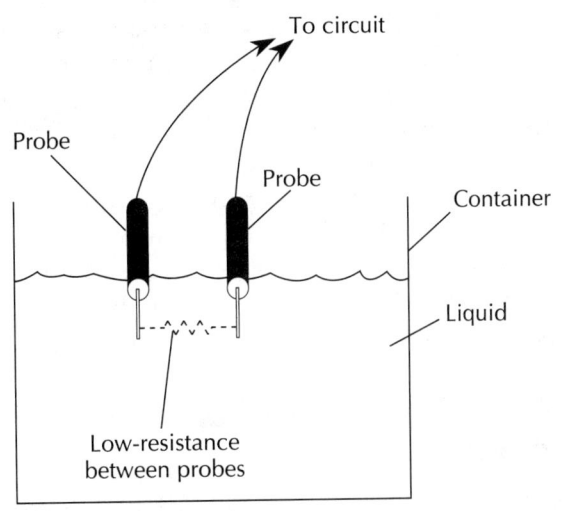

Fig. 3-2 *When liquid touches both probes, the liquid functions like a resistor connected between the two separated probes.*

The exact frequency change will depend on the specific liquid involved, because different liquids have different resistances. If ordinary water is detected, the frequency will almost double when the probes are immersed. Perhaps you will be able to identify the type of liquid just from the change in the pitch generated by the circuit.

In many applications, it might be annoying or otherwise impractical to have the tone continuously sounding. One possible solution would be to increase the value of capacitor C1. This reduces the base (dry probes) frequency of the oscillator. If this base frequency is set just below the point of audibility (about 20 to 30 hertz), when liquid is sensed between the probes, the frequency will jump up into the audible region. Unfortunately, the liquid sensing tone in this case would have a very low pitch, which can sometimes be hard to hear.

Another solution would be to add a normally open SPST push-button switch between the supply voltage (a 1.5 V battery cell) and pin 5 of the LM3909. This way, the circuit will operate, only when the push button is being held closed. You can design the enclosure of this project so that the push button can easily be held down while it is in use.

For use as a pouring aid, the probes of this circuit should be mounted in the container at the desired maximum level for the liquid. When the liquid being poured into the container reaches this level, it electrically connects the probes, causing the tone to rise in pitch. It should not be difficult to construct a mechanical probe holder that fits neatly over the rim of a cup or glass.

As with any electronic project that comes into contact with water, you should use battery power only. Never use an ac power source with this type of project. If the monitored liquid gets into the circuitry itself, it could cause a potentially dangerous short circuit. Why take a risk of a fire or a painful, or even fatal electrical shock? There is no need for ac power because batteries will operate this circuit for such a long time before they need replacement.

Moisture detector

The circuit shown in Fig. 3-3 is designed to detect a small amount of liquid, such as from a small leak or spill. For greater amounts of liquid, it would be better to use one of the flood detector-projects described in this chapter.

A suitable parts list for this project is given in Table 3-2. None of the component values in this circuit are particularly critical. Any convenient substitutions that are reasonably close to the suggested values should work in this project.

The sensor is made up of a mesh of fine wires spaced an inch or so apart. The wires of the two halves of the sensor probe are interleaved. The basic pattern is shown in the unusual symbol used for the sensor in the schematic diagram. You can't buy this type of sensor. You'll have to make it yourself. It shouldn't take you more than a few minutes, and the cost is just a few pennies worth of ordinary hook-up wire. Stranded wire would be slightly better than solid wire for this application, although either will work. If you use stranded wire, be careful not to let any loose strands touch the other half of the sensor. If the strands touch, they will create a short circuit.

An alternate approach to the sensor is a small printed circuit board etched with a similar interleaved pattern, as shown in Fig. 3-4.

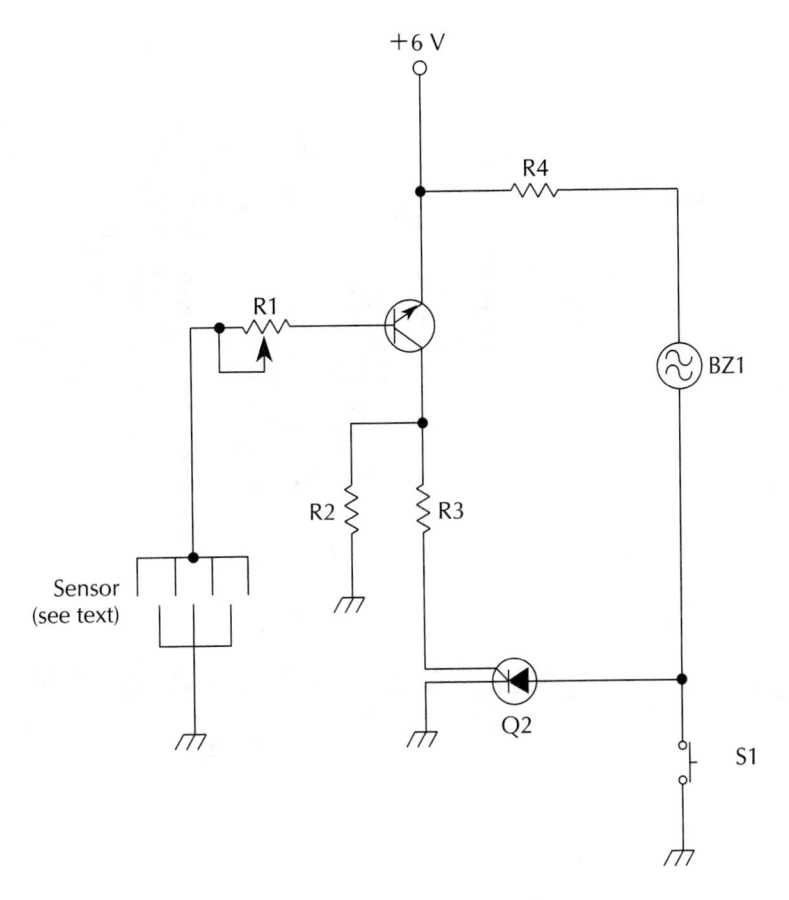

Fig. 3-3 *Project 14. Moisture detector.*

Table 3-2 Suggested parts list for Project 14. Moisture detector.

Q1	PNP transistor (2N3906 or similar)
Q2	low-power SCR to suit load
BZ1	alarm sounder (buzzer, bell, etc.—see text)
S1	normally open push-button switch
R1	100 kΩ potentiometer
R2	10 kΩ ¼ W 5% resistor
R3	1 kΩ ¼ W 5% resistor

Virtually any PNP transistor will work for Q1 in this circuit, and Q2 can be any low-power SCR. The exact specifications of these components is not critical to the operation of this project. They form a simple electronic switch. Normally, the sensor,

104 Liquid-related projects

Interleaved probe

Etched area

Interleaved probe

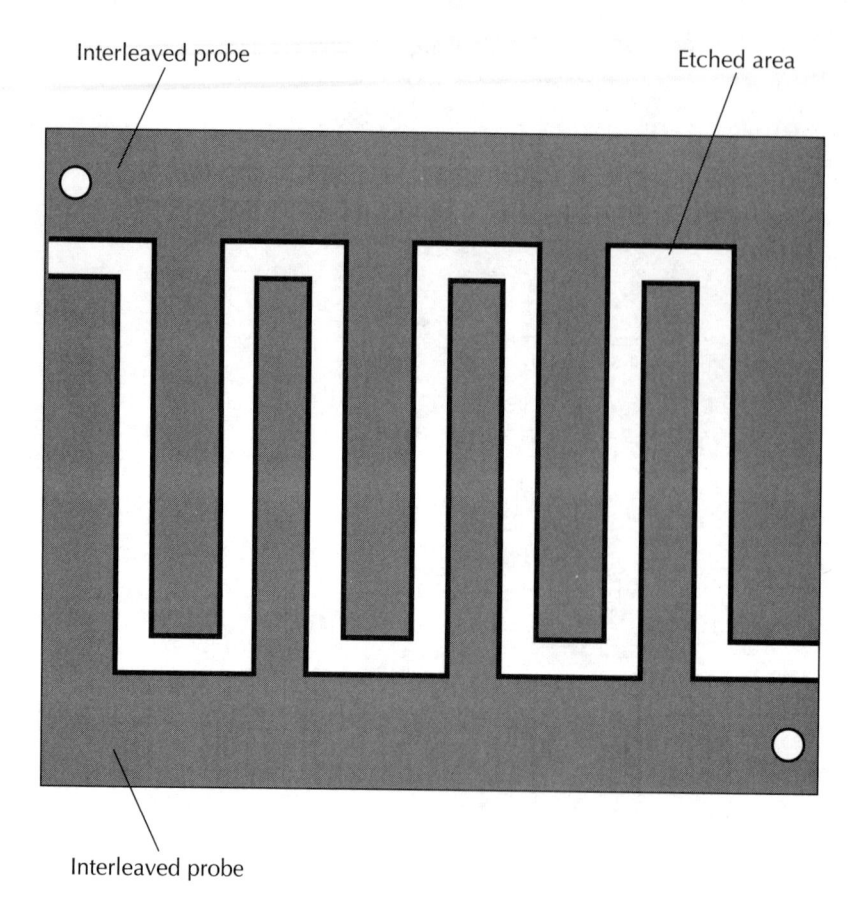

Fig. 3-4 *A small printed circuit board etched with an interleaved pattern can be used as the moisture sensor for Project 14.*

when dry, exhibits a high resistance from the base of transistor Q1 to ground. When the sensor gets wet, the moisture will create a fairly low resistance across the two halves of the sensor. This will trigger the electronic switch, firing the SCR and causing the buzzer to sound.

Temporarily close normally open push-button switch S1 to silence the alarm and reset the circuit.

For some combinations of Q1 and Q3, you might find it necessary to experiment with somewhat different values for resistor R3. It would be a very good idea to breadboard this circuit on a solderless breadboard and check it out before you heat up your soldering iron.

Trimpot R1 permits you to calibrate the sensitivity of your moisture detector project. Just adjust the resistance of this trimpot until the buzzer sounds reliably when the sensor is wet, but not when it is dry. In some applications, you might prefer to replace this trimpot with a fixed resistor with a value of about 100 kΩ.

The sensor should be separate from the main circuit board, connected by a moderately long pair of wires. You probably won't need shielding for these leads in any practical environment. Place the sensor where it will get wet as soon as possible in the event of a problem to be detected. The rest of the circuit should be kept where it can be kept dry. If the main circuit board gets wet, you will get all sorts of short circuit problems, and the transistor and SCR are likely to be damaged or destroyed.

You might want to experiment with different values for resistor R4 to control the volume of the alarm sound. The larger this resistor is, the softer the buzzer sound will be. In some cases, you might want to eliminate this resistor from the circuit altogether for the maximum alarm volume.

If you prefer, you can replace the buzzer with some other alarm device, such as a bell or electronic siren or tone generator of some sort. It would probably be a good idea to delete resistor R4 in most such cases.

The output of this circuit can even be used to drive other circuitry, activating some sort of automation device to respond to the problem directly, rather than just crying for help. For example, if the moisture detector determines that a pipe is leaking, it could trigger an automated circuit to shut off the water flow through that pipe before any serious damage can be done. Use your imagination.

As with any electronic project that comes into contact with water, you should use battery power only. Never use an ac power source with this type of project. If the monitored liquid gets into the circuitry itself, it could cause a potentially dangerous short circuit. Why take a risk of a fire or a painful, or even fatal electrical shock? There is no need for ac power because batteries will operate this circuit for such a long time before they need replacement.

Simple plant-watering monitor

Some people seem to have a green thumb and can keep their house plants living with very little apparent effort. Other people seem to have a *black thumb*—almost every plant they get a hold of quickly withers and dies. Frequently the chief problem in dying house plants is the owner simply forgot to water them enough. Or perhaps she or he gave them too much water. Either extreme can be disastrous for a plant's health.

This project can automatically keep track of the moisture in the soil your plants are in and remind you when it is time to add more water. The schematic diagram for the simple plant-watering monitor is shown in Fig. 3-5. A suitable parts list for this project appears in Table 3-3.

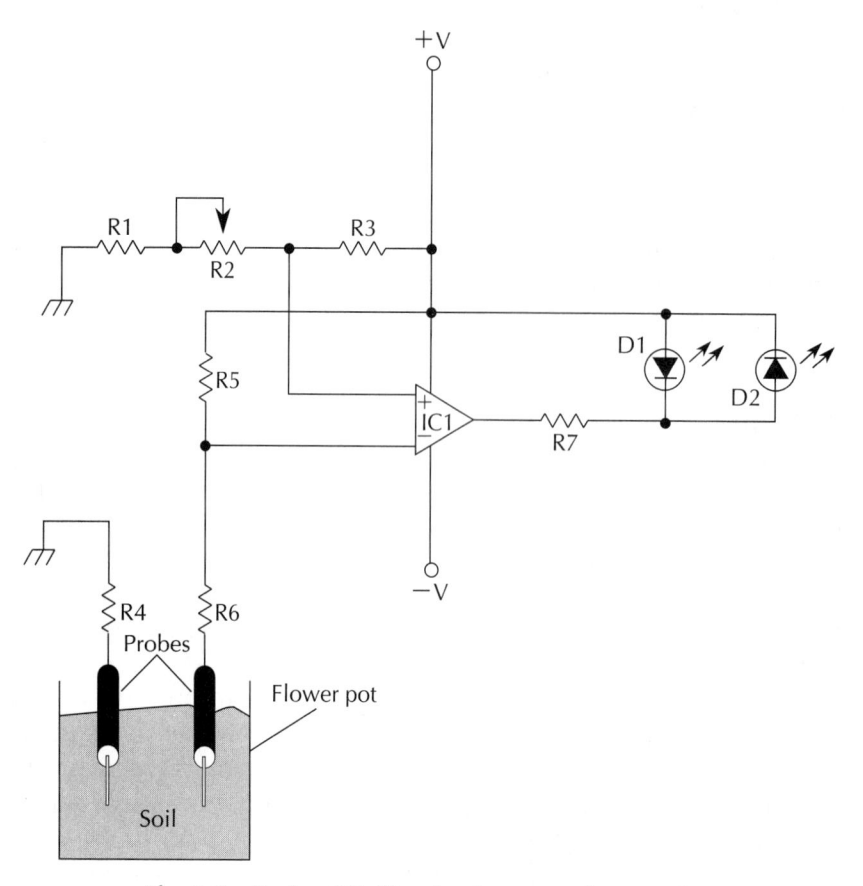

Fig. 3-5 *Project 15. Simple plant-watering monitor.*

Nothing is particularly critical here. Almost any op amp should work well in this circuit. There would be no real advantage in spending extra money on a high-grade, low-noise op amp chip in this application.

As in many water sensing circuits, the sensor probes are just a pair of metallic probes (stiff wires or metal rods), connected at one end to the circuit via a pair of simple test leads. The other ends of the probes are left free. The free ends of both probes are

Table 3-3 Suggested parts list for Project 15. Simple plant-watering monitor.

IC1	Op amp (741 or similar)
D1, D2	LED
R1, R3, R4, R5	100 kΩ ¼ W 5% resistor
R2	1 MΩ potentiometer
R6	1 kΩ ¼ W 5% resistor
R7	330 Ω ¼ W 5% resistor

inserted into the soil to be monitored. In most cases, they should be placed about one to two inches apart. Do not place the probes too much farther apart than this, or the project might not work reliably. For best results, bury the probes near the roots of the plant you want to monitor. But be careful you don't get too close and possibly damage the roots with the probes. This is especially important for delicate plants.

It might seem too obvious to mention, but the author has received a similar question on a project in one of his previous books. So the answer is repeated here. If you are monitoring house plants, both probes must be in the same pot. Just because there are two probes, that doesn't mean that you can simultaneously monitor two separate potted plants with one circuit. The two probes must work together.

In operation, the circuitry senses the resistance between the two probes. Dry soil will exhibit a high resistance between the probes. The more moisture there is in the monitored soil, the lower the resistance (or the higher the conductivity).

In effect, the soil between the two probes acts like a variable resistor, which forms part of a voltage-divider string along with actual resistors R4, R5, and R6. To make this clearer, this part of the circuit is shown separately in Fig. 3-6.

A second voltage-divider string is made up of R1, R2, and R3. The op amp (IC1) acts as a simple voltage comparator. Its output indicates which of the two voltages (from the tap off point in the two voltage divider strings) is the higher.

R2 is a manual potentiometer. The potentiometer permits you to manually change the reference voltage seen by the comparator (IC1). The monitor voltage is controlled by the only variable resistance element in the lower voltage divider string—the resistance of the soil between the two probes. All other resistances in both voltage-divider strings are fixed, and you can ignore them in this theoretical discussion, because their effects are constant.

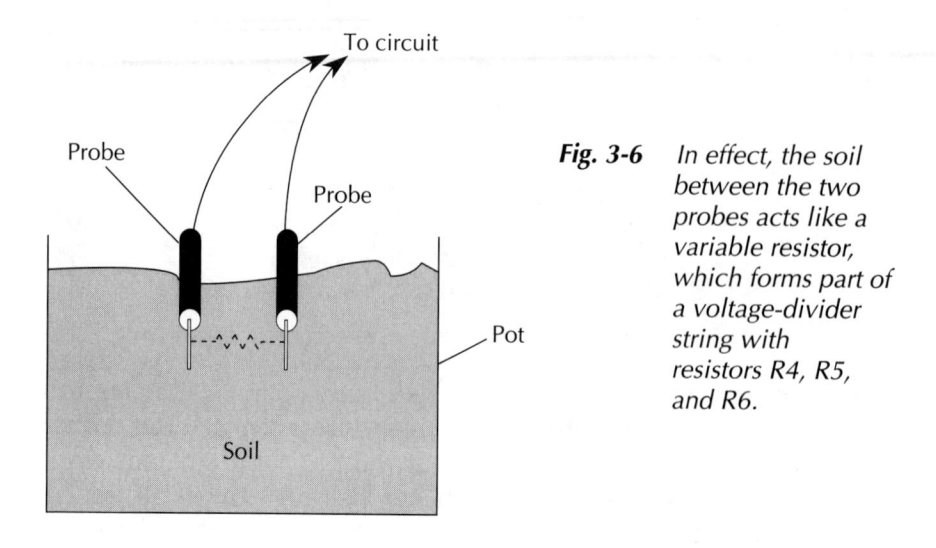

Fig. 3-6 *In effect, the soil between the two probes acts like a variable resistor, which forms part of a voltage-divider string with resistors R4, R5, and R6.*

A trimpot should be used for potentiometer R2, rather than a front-panel control, as this is a calibration control. Once it is properly set, leave it alone.

Because the resistance of R2 is not changed once it has been set, the only variable in the input circuitry (as seen by the comparator (IC1)) is the resistance between the two probes. Ignoring the other resistances in the circuit, you can say that the soil is too dry when its resistance is less than that of R2, which will mean the probe voltage is higher than the R2 (reference) voltage, and the comparator output will be triggered.

When the soil is sufficiently moist, one LED will be lit. If it gets too dry the other LED will light up. Resistor R7 is a current-limiting resistor for both LEDs. For the best visibility, it is strongly recommended that you use LEDs with two different colors. Red and green are good choices. They are both widely available, and are very easy to see and distinguish.

To calibrate the project, make sure that the soil is almost, but not quite, too dry. That is, when the plant is due for a regular watering is the time to perform the calibration procedure. Of course, you must wait to water the plant until after the calibration is completed. If you're not sure about how to determine the correct soil condition, ask a friend who is good with plants to help you. Adjust potentiometer R2 until the too-dry LED comes on, then back off, until the moist LED just barely comes on. It might be a little tricky to get this set exactly right. A ten-turn trimpot would probably help. On the other hand, this application isn't all that critical. If you can adjust a standard single-turn trimpot so it is reasonably close to the true calibration point, the project should work well enough.

You should be aware that different types of soil will have different resistances, and different soils will require recalibration. Also, some plants need more moisture, and others do better in somewhat drier soil. Again, recalibration might be required if you are monitoring several different types of plants. Because this circuit is so simple and inexpensive, it would probably be a lot less fuss and bother to build a separate copy of the project for each plant you want to monitor.

Notice that this project requires a dual power supply (V+ and V−, both referenced to ground) for most inexpensive op amp chips. Some op amp devices can be operated from a single-ended power supply. A dedicated comparator chip like the LM339 should also work fine. If in doubt about the power supply requirements, check the manufacturer's specification sheet for this information.

The actual supply voltage(s) is not critical to the operation of this project. The primary determinant is the power requirements of the op amp itself. Check the specifications sheet. Most modern op amp devices can be operated over a wide range of supply voltages. Of course, when a dual-polarity power supply is used, the V+ and V− voltages should be equal, except for their opposite polarities. As long as the op amp is fed an acceptable supply voltage, raising or lowering the supply voltage (within the rated range) should make no difference at all in the operation of the circuit.

If you prefer, you could easily modify this project to trigger a buzzer or some other audible alarm when the monitored soil gets too dry. Some people might find this helpful if they tend to neglect to even look at the LEDs for extended periods. Others might find an audible alarm extremely annoying—especially if it goes off in the middle of the night. As with so many electronic projects, the decision is yours. It's a matter of individual preference and the particular requirements of your specific intended application.

As with any electronic project that comes into contact with water, you should use battery power only. Never use an ac power source with this type of project. If the monitored liquid gets into the circuitry itself, it could cause a potentially dangerous short circuit. Why take a risk of a fire or a painful, or even fatal electrical shock? There is no need for ac power because batteries will operate this circuit for such a long time before they need replacement.

Deluxe plant-watering monitor

For true gardening enthusiasts, the plant-watering monitor project described above won't be of much use. Gardening enthusiasts have already trained themselves to know when the soil is dry and when it is moist. When working with delicate plants, especially hybrids, more precise information about soil moisture content might be required. Instead of a simple yes/no monitor like the last project, such gardening enthusiasts could use an actual meter that gives them specific measurement values.

A circuit for accomplishing this is shown in Fig. 3-7. A suitable parts list for this project is given in Table 3-4.

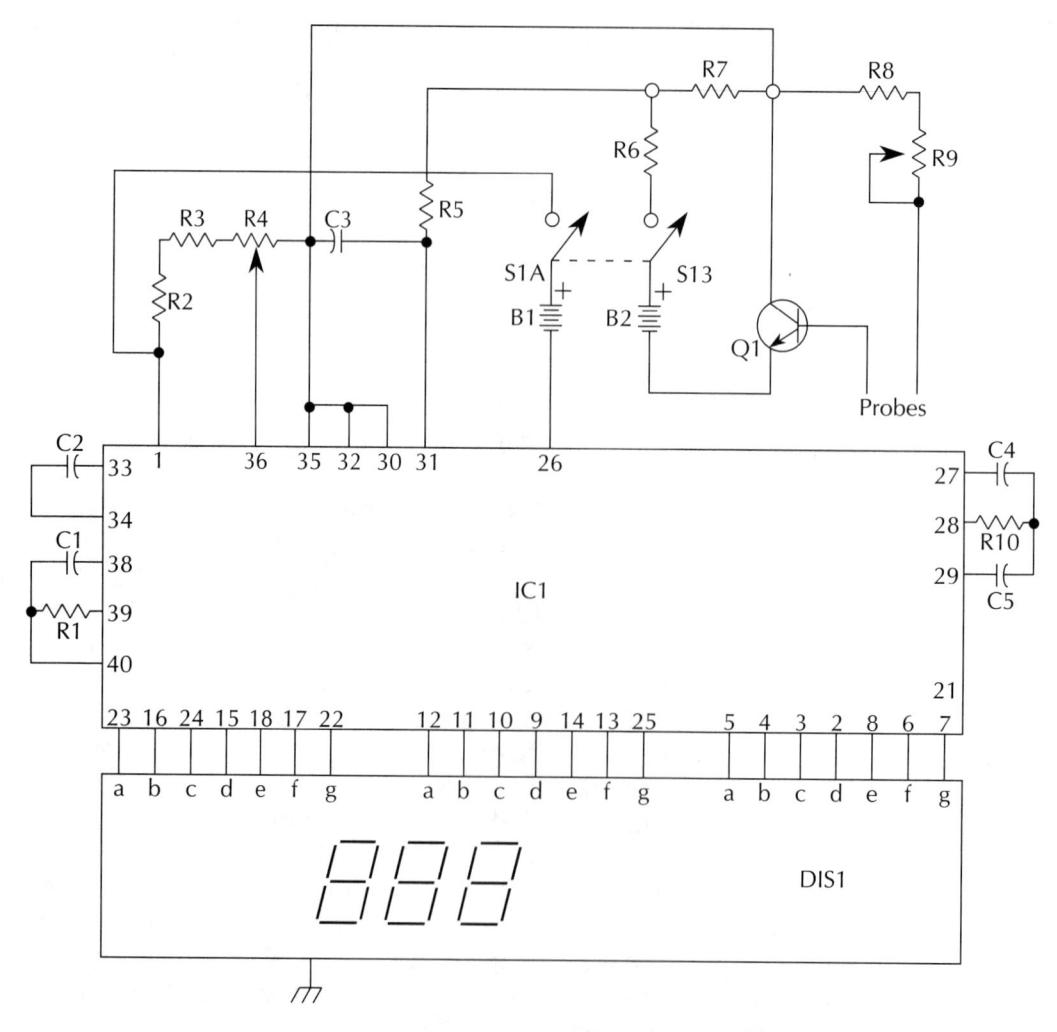

Fig. 3-7 *Project 16. Deluxe plant-watering monitor.*

Table 3-4 Suggested parts list for Project 16. Deluxe plant-watering monitor.

IC1	ICL7106 A/D converter/LCD driver (intersil)—see text
Q1	NPN transistor (2N3904 or similar)
DIS1	Three-digit LCD display unit
C1	100 pF capacitor
C2	0.1 µF Mylar capacitor
C3	0.01 µF Mylar capacitor
C4	0.047 µF Mylar capacitor
C5	0.22 µF Mylar capacitor
B1	9 V battery
B2	3 V battery
R1, R8	10 kΩ ¼ W 5% resistor
R2, R6	2.2 kΩ ¼ W 5% resistor
R3	22 kΩ ¼ W 5% resistor
R4	25 kΩ trimpot
R5	1 MΩ ¼ W 5% resistor
R7	100 Ω ¼ W 5% resistor
R9	1 MΩ trimpot
R10	470 kΩ ¼ W 5% resistor

The projects in this book call for commonly available, general-purpose components as much as possible, to try to avoid the frustration some readers might face when they can't find a critical component to complete their project. For this project, however, a specialized component is necessary. Before bothering with anything else about this project, make sure you can find a source for IC1. This is a rather unusual and specialized device. It is the ICL7106 A/D (analog-to-digital) converter and LCD driver IC. The ICL7106 is not uncommon, but of course, there is no way of guaranteeing it will remain available in the future. Specialized electronic components are developed or become obsolete very quickly and without much advance warning these days. Because of the complexity of this type of chip, substituting a different IC designed for similar functions probably wouldn't be worth the effort. The pin numbers and the required support circuitry is likely to be so different, that you will essentially have to redesign the entire project from scratch.

On the other hand, trying to duplicate the functions of the ICL7106 using discrete, general-purpose components, although possible, is highly impractical. The circuit would be physically large and unwieldy, and the project would also be ridiculously expensive. So a dedicated, specialized IC is really the only reasonable way to go in this project.

The ICL7106 is essentially a complete digital voltmeter on a single chip. It converts the analog input voltage into appropriate digital values via the A/D converter, then it conditions the output signals to drive a three digit seven segment LCD (liquid-crystal display) unit. The LCD unit (DIS1) is referenced to ground. Check the required connections on the manufacturer's specification sheet. These connections will vary considerably from unit to unit, so including pin numbers for DIS1 on the schematic diagram would be almost useless.

Mylar or polystyrene capacitors are strongly recommended for all capacitors throughout the circuit, except for C1. Common ceramic disc capacitors won't really be accurate enough for this application. Similarly, use resistors rated for tolerances no more than 5% in this project. Resistors with 10% or 20% accuracy just won't be good enough in this project.

Almost any low-to-medium power NPN transistor should work okay for Q1. Any minor differences from using a different transistor type number should be compensated by the adjustment of full-scale calibration trimpot R9.

Notice that this project requires two power supply voltages—+9 V and +3 V. A standard 9 V transistor battery and a pair of 1.5 V AA or AAA batteries would be a good choice, making the project nicely compact and portable. The power consumption of this circuit is quite low, so the batteries should last quite a long time before needing replacement. Typically just a little more than 2 mA will be drawn from the 9 V battery. The current drain on the 3 V battery (two 1.5 V batteries in series) will be less than 300 µA, which is practically negligible.

Two trimpots (R4 and R9) are used to calibrate this project. Ten-turn trimpots will permit more accurate calibration, of course, but they will tend to be considerably more expensive. Whether the increased cost is worth it will depend on your individual intended application.

The calibration procedure must be completed before inserting the monitor probes into the soil to be monitored.

To begin the calibration procedure, leave the probes unconnected to anything. Make sure they are not touching. This simulates the maximum possible degree of dryness. Adjust trimpot R4 for a reading of 000. This should be near the centerpoint of the trimpot range. If it is not, you might want to experiment with slightly different values for resistor R2. This will rarely be absolutely essential, but it might make fine tuning the calibration of the unit a little easier.

Now place the tips of the probes into a glass of water. Of course, this simulates the maximum possible moisture level. Adjust trimpot R9 for a full-scale reading of 100.

Remove the probes from the water and let them dry off. Repeat the zero test described above. You might find you have to readjust trimpot R4 a little. If so, repeat the water test and readjust R8, if necessary for a reading of 100. Go back and forth between the two calibration extremes, until both the 000 and the 100 readings are stable. Again, how picky you want to get about this depends on how critical your application is. If you aren't too fussy about exactly perfect results, you might consider a little *bobble* of the digits at the extreme ends of the monitor range acceptable.

Once calibrated, the project will display the relative moisture content sensed by the probes in percent, ranging from 000% (perfectly dry) up to 100% (totally wet). A good gardening guide should tell you how much relative moisture is suitable for making various particular types of plants thrive the best. You will no longer have to worry about the possibility of over-watering or under-watering your expensive, delicate plants.

As with any electronic project that comes into contact with water, you should use battery power only. Never use an ac power source with this type of project. If the monitored liquid gets into the circuitry itself, it could cause a potentially dangerous short circuit. Why take a risk of a fire or a painful, or even fatal electrical shock? There is no need for ac power because batteries will operate this circuit for such a long time before they need replacement.

Flooding alarm

Do you ever have a problem with occasional water leakage causing minor flooding in your basement? Do you need some sort of automated indication when a tank (or other container) of liquid is filled past a specific point? This simple flooding alarm project will let you know whenever conditions of this type occur.

The schematic diagram for the flooding alarm circuit is shown in Fig. 3-8. The suggested parts list for this project appears in Table 3-5.

As you can see, this is a very simple project. Most of the work is done by IC1, which is an LM3909 LED flasher/oscillator chip. This device is quite widely available, and it should continue to be available for some time to come. It is a very simple

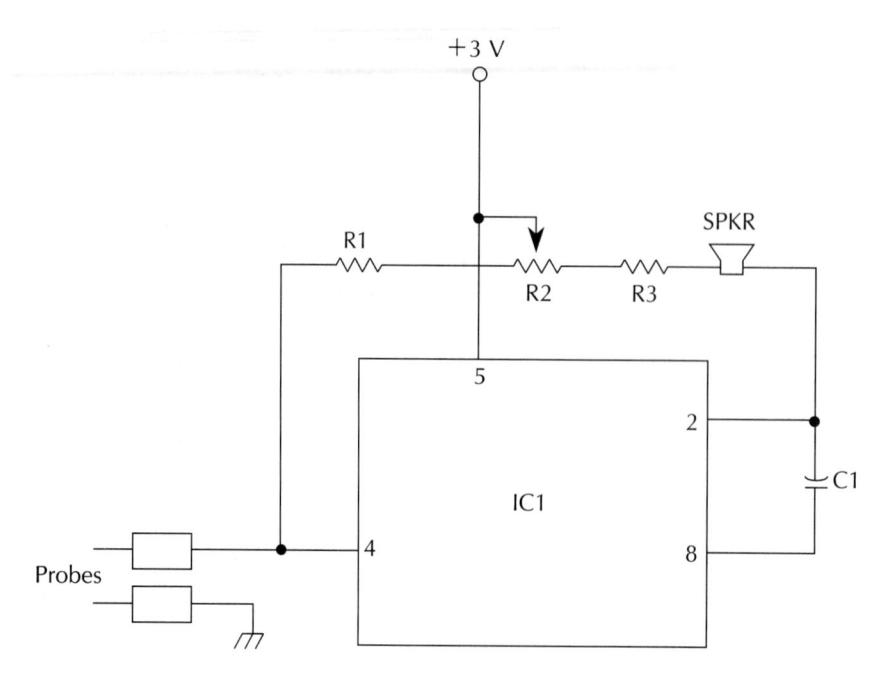

Fig. 3-8 *Project 17. Flooding alarm.*

Table 3-5 Suggested parts list for Project 17. Flooding alarm.

IC1	LM3909 LED flasher/oscillator
SPKR	small 8 Ω speaker
C1	0.1 µF capacitor
R1	1 MΩ ¼ W 5% resistor
R2	100 Ω potentiometer
R3	22 Ω ¼ W 5% resistor

device with a surprisingly wide number of applications. In this project the LM3909 is used in its high frequency mode, in which it acts like a simple audio oscillator.

None of the component values in this circuit are particularly critical. The basic frequency of the tone is determined primarily by the value of capacitor C1. You might want to try experimenting with alternate values for this capacitor to achieve the most pleasing effect.

Potentiometer R2 acts as a volume control for the alarm. Resistor R3 is included in the circuit to prevent the volume control from being set to 0, possibly overloading and damaging the speaker. If a manual volume control is not needed in your particular intended application, you can replace R2 and R3 with a single small-value fixed resistor.

When the probes are dry, there is a very high resistance, so pin 4 of the LM3909 will not be grounded, and the oscillator circuit is inhibited from functioning. It just sits there and waits.

To prevent possible false alarms, a small positive voltage is applied to the IC ground pin (pin 4) through pull-up resistor R1. The exact value of this pull-up resistor is not critical, although it should be fairly large, so it won't confuse the circuitry when water is detected between the probes.

When the water level rises, making contacts with both probes, they are effectively shorted together through the relatively low resistance of the liquid. At this point, there will be a fairly low resistance between pin #4 of IC1 and ground. This resistance to ground will be considerably less than the resistance of the pull-up resistor (R1). The LM3909 is now properly grounded, which permits it to function, and the tone is sounded through the speaker. When you hear the tone, you know a flooding condition has been detected and should be corrected. The exact resistance between the probes during flooding will depend on the specific liquid involved. Some liquids have more resistance than others. The value of pull-up resistor R1 must be considerably higher than the resistance of the liquid between the circuit probes, or the project will not work. You might have to do some experimentation here. If you are only interested in detecting ordinary water, you should have no problem, because water has a very low resistance, and the component values suggested in the parts list will be more than adequate.

The probes of this flooding alarm should be mounted slightly below the truly dangerous or damaging level, so you will have a chance to attempt to correct the problem before it gets too bad. For example, if 15 inches of water will cause major problems, it would be advisable to mount the probes at the 12- or 13-inch level, instead of at the critical 15-inch position.

But what if you don't want any standing water at all? In this case, mount the probes flat on the floor. As soon as there is enough liquid present to cover both probes, the alarm should sound.

As with any electronic project that comes into contact with water, you should use battery power only. Never use an ac power source with this type of project. If the monitored liquid gets into the circuitry itself, it could cause a potentially dangerous short circuit. Why take a risk of a fire or a painful, or even fatal electrical shock? There is no need for ac power because batteries will operate this circuit for such a long time before they need replacement.

Visual flooding alarm

In a few special applications, you might find you need a flooding alarm that produces a visual signal instead of an audible alarm. Fig. 3-9 shows the schematic diagram for a simple, but effective flooding alarm circuit. A suitable parts list for this project is given in Table 3-6.

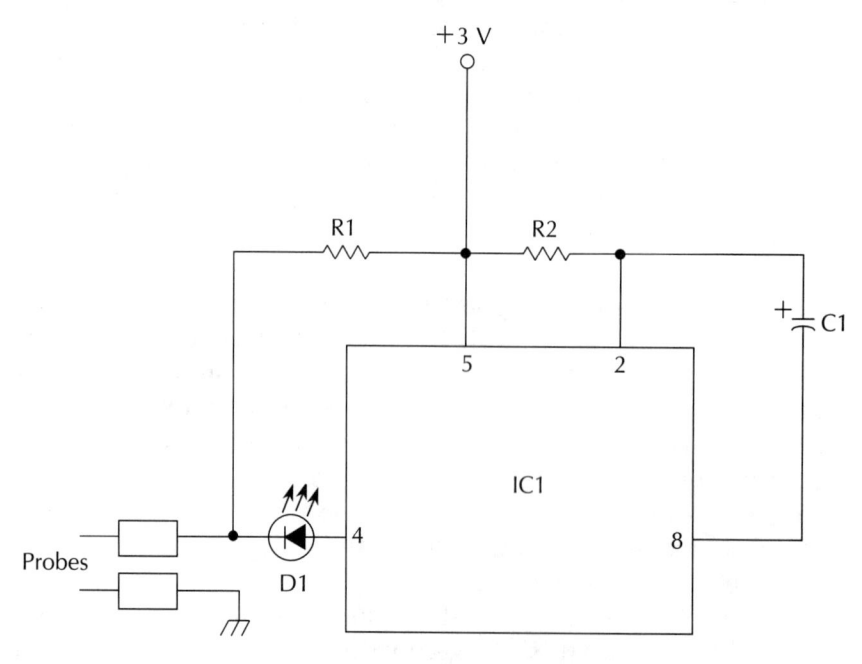

Fig. 3-9 Project 18. Visual flooding alarm.

Notice that once again, this project is built around the LM3909 LED flasher/oscillator IC. This time this chip is being used in its low-frequency LED flasher mode. Otherwise, this circuit is basically the same as in the previous project.

When the oscillator in this project is operating, it produces a square wave with a frequency that is too low to be heard through a speaker. If you apply this low-frequency output signal across the LED, you will be able to see it flash on and off in a regular pattern. A flashing LED is easier to see and more eye catching than a continuously lit LED.

Table 3-6 Suggested parts list for Project 18. Visual flooding alarm.

IC1	LM3909 LED flasher/oscillator
D1	LED
C1	220 µF 10 V electrolytic capacitor
R1	1 MΩ ¼ W 5% resistor
R2	100 Ω ¼ W 5% resistor

Normally, when the circuit probes are dry, the LED (D1) remains dark. The circuit just sits there, patiently waiting. It doesn't even consume any current from the battery in this project. As in the previous audible flooding alarm project, the high resistance between the dry probes is essentially an open circuit. There is no ground connection to pin 4 of the LM3909 (IC1). This lack of ground prevents the chip from operating. This is the same as turning off a power switch. Resistor R1 is a pull-up resistor to prevent the possibility of any false alarms, as in the preceding project. A small positive voltage is applied to the IC ground pin (pin 4) through this pull-up resistor R1. The exact value of this pull-up resistor is not critical, although it should be fairly large, so it won't confuse the circuitry when liquid is detected between the probes.

When the water level rises, making contacts with both probes, they are effectively shorted together through the relatively low resistance of the liquid. At this point, there will be a fairly low resistance between pin 4 of IC1 and ground. This resistance to ground will be considerably less than the resistance of the pull-up resistor (R1). The LM3909 is now properly grounded, which permits it to function, and the LED (D1) will start flashing. As it flashes, the alarm alerts you to take action to correct the flooding condition, before serious damage is done.

The LED flash rate is determined primarily by the value of capacitor C1. You might want to experiment with alternate values of this capacitor. Don't make the capacitor too small in value, however. If the LED flash rate is increased above about 10 Hz (hertz) or so, your eye won't be able to distinguish between the individual blinks. They will all blur together, and the LED will appear to be continuously lit. This phenomenon is known as persistence of vision, and it is the principle that lets you see movies as moving images, rather than a sequence of still pictures (which is what a movie really is). If the flash frequency is too high, the LED will appear to be continuously lit, although perhaps at slightly less than full brightness. It is still really blinking on and off, but it is too fast to see.

As in the previous audible flooding alarm project, the probes of this visual flooding alarm circuit should be mounted slightly below the truly dangerous or damaging level, so you will have a chance to attempt to correct the problem before it gets too bad. For example, if 3 inches of water will cause major problems, it would be advisable to mount the probes at the 1.5- or 2-inch level, instead of at the critical 3-inch level.

But what if you don't want any standing water at all? In this case, mount the probes flat on the floor. As soon as there is enough liquid present to cover both probes, the alarm will be triggered.

The LM3909 LED flasher/oscillator IC happens to have exceptionally low power-supply requirements. The current drain presented to the battery by the LM3909 is virtually negligible. It is actually little more than the normal leakage current of a battery sitting unused on a shelf. In other words, the battery life in a typical LM3909 circuit under continuous operation would be almost as long as the ordinary shelf life of the battery. Of course, in the quiescent mode (when the LED is dark), the circuit will draw no current from the battery at all, because the IC is turned off by the open circuit between the probes. The LED (D1) is the big consumer of current in this project, and the current is internally limited in the LM3909 to a much lower than usual level.

As with any electronic project that comes into contact with water, you should use battery power only. Never use an ac power source with this type of project. If the monitored liquid gets into the circuitry itself, it could cause a potentially dangerous short circuit. Why take a risk of a fire or a painful, or even fatal electrical shock? There is no need for ac power because batteries will operate this circuit for such a long time before they need replacement.

Only the probes should be exposed directly to the liquid being monitored by the project. The probes should be connected at the end of fairly long lead wires, and not connected to the main circuit board itself. That would just be asking for trouble. Mount the main body of the project where it is least likely to get wet. But bear in mind, a flooding condition could be worse than you expect—don't use ac power for a flooding alarm project, no matter how cleverly you have it mounted. There is always the possibility of something totally unforeseen going wrong.

In some applications, you might want to use this project backwards, to warn you when the liquid in some monitored container has fallen below some critical level. In this case, the LED should normally be flashing. When it goes dark, you know you have a problem to attend to before it creates too much trouble.

Sump-pump controller

A flood alarm that alerts you when flooding is taking place is a good idea, and you can use it to prevent hundreds of dollars of potential damage to your property. But suppose no one is home when the flooding occurs. The alarm sounding won't do any good at all. For even better security, you might want to combine a flooding detector with some sort of automation system that can actively respond to correct the situation, even when no one is home when the problem strikes.

One of the primary applications for a flooding detector is to control a sump pump to automatically correct the problem as soon as it is detected. When the circuit detects flooding, the sump pump starts running, until enough of the water has been removed, then the pump is shut down. Keeping the pump running continuously would waste a lot of energy, and would put a lot of unnecessary wear and tear on the pump itself. Many pump designs can be damaged by being run dry for an extended period.

A simple, but effective automated sump-pump controller circuit is shown in Fig. 3-10. The parts list for this project is given in Table 3-7.

Notice that this circuit has three separate probes. The correct relative placement of these probes is critical to the operation of the circuit. The off probe should be mounted somewhat lower than the on probe. This will prevent hysteresis problems that could cause the sump pump to repeatedly cycle on and off without really accomplishing very much. The common probe should be lined up with the off probe, or a little lower.

If the water in the monitored area rises enough to touch the common and off probes, nothing happens. However, when the water reaches the on probe, the circuit relays are triggered, and the sump pump is turned on.

Unless something is disastrously wrong, running the pump will lower the water level in the area. Eventually the level will drop below the on probe, but nothing will happen at this point. The sump pump will continue to run, removing more water. It is not until the water level drops below the off probe that the sys-

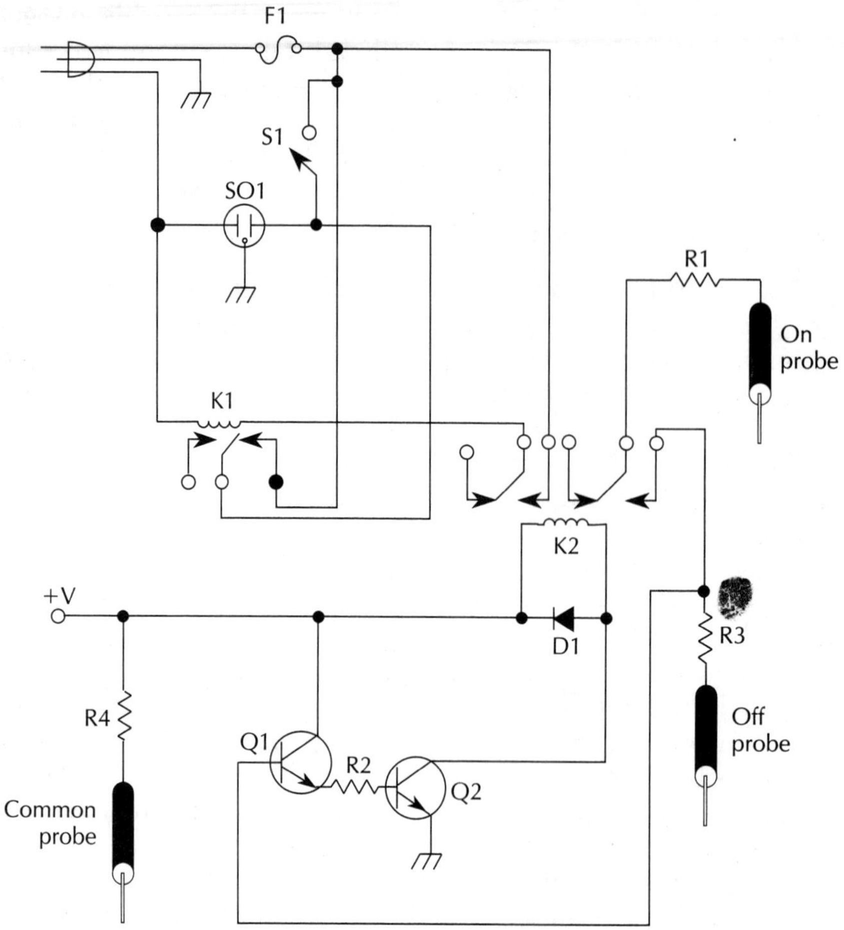

120 Liquid-related projects

Fig. 3-10 Project 19. Sump-pump controller.

**Table 3-7 Suggested parts list
for Project 19. Sump-pump controller.**

Q1, Q2	NPN transistor (2N3904 or similar)
D1	diode (1N4002 or similar)
K1	120 Vac SPST relay (contacts to suit load—pump)
K2	DPDT dc relay
F1	fuse (to suit load—pump)
S1	SPST switch
R1–R4	1 kΩ ¼ W 5% resistor

tem will shut down. This prevents the sump pump from oscillating on and off if the water level is just at the edge of the on probe.

In most applications, the off probe should be mounted as low as physically possible, perhaps even lying flat on the floor itself. The on probe should be at least a couple inches above the floor, but sufficiently low that no major flooding damage can be done before the sump pump is automatically turned on by the circuit.

As with any electronic project that comes into contact with water, you should use battery power only. Never use an ac power source with this type of project. If the monitored liquid gets into the circuitry itself, it could cause a potentially dangerous short circuit. Why take a risk of a fire or a painful, or even fatal electrical shock? There is no need for ac power because batteries will operate this circuit for such a long time before they need replacement.

Water-heater controller

If you have an oil-based water heater, this project could save you quite a bit on fuel bills. In water heaters fueled by oil, the system water temperature is controlled by a device called an *aquastat*. Usually there are manual adjustments for water temperature and circulator control. In most cases, the circulator is set about 20°F lower than the water temperature.

Greater fuel efficiency can be achieved by using a somewhat lower water temperature setting during warmer weather than in colder weather. Good rule-of-thumb values are about 180°F for the winter months and about 160°F in the summer. You are assuming that the interior of the home is heated in the winter and cooled during the summer, so the indoor room temperature will be more or less consistent. You aren't so much interested in the outside temperature as an absolute measurement as with the difference between the indoor and outdoor temperatures. Notice that with these recommended values, the water temperature varies inversely with the outside temperature. When it gets colder outside, the water should be heated more, and vice versa.

Of course, it is all too easy to forget to make these seasonal changes in the water heater setting. Also, the weather in many areas can vary over quite a wide temperature range at certain times of the year. An efficient water temperature setting one day might be quite wasteful the next day. Also, remember that the outside temperature tends to change quite a bit between the day and night. It would obviously be highly impractical to manually readjust the water temperature setting on your water heater each time the weather changes. This is a very good application for electronic automation. The heater controller circuit presented here can automatically make the necessary adjustments in the water temperature to maximize your fuel efficiency, and give you some impressive savings on your fuel bill over time.

This project is designed to monitor the outside and indoor temperatures and automatically adjust the water temperature according to the detected difference. For best results, the circulator control on your heater should be set to about 125 to 135°F. Because this circuit takes direct measurements of the actual temperature, you don't need to bother to make seasonal adjustments of your heater water temperature control. About the only time you'll need to adjust the controls will be if you expect to be away from home for an extended period. There is no point in using fuel to heat water that's just going to sit in the heater and not be used. You will probably want to keep the heater running at a much lower temperature, however, especially during the winter months in very cold areas. At the very least the heater temperature should be high enough to prevent the possibility of freezing.

Of course, it will also probably be necessary to readjust the controls whenever the heater needs servicing. But in general, everyday use, you can set the controls once and forget about them. The heater controller project will keep track of things from then on.

(It would be a good idea to get in the habit of manually checking out the system periodically to catch anything that might possibly go wrong with the controller circuit, or the heater itself.)

The schematic diagram for the heater controller circuit is shown in Fig. 3-11. A suitable parts list for this project appears as Table 3-8.

Nothing is particularly critical in this project. The parts list calls for only simple, commonly available components, and there is plenty of leeway throughout the circuit for making reasonable substitutions.

The heater aquastat is controlled by the switch contacts of the relay, which serves as the output of this project.

The heart of this circuit is IC1, a common op amp chip. Any standard op amp device should work fine in this application. There would be no particular advantage to using a high-grade, low-noise op amp chip in this circuit. A common, inexpensive

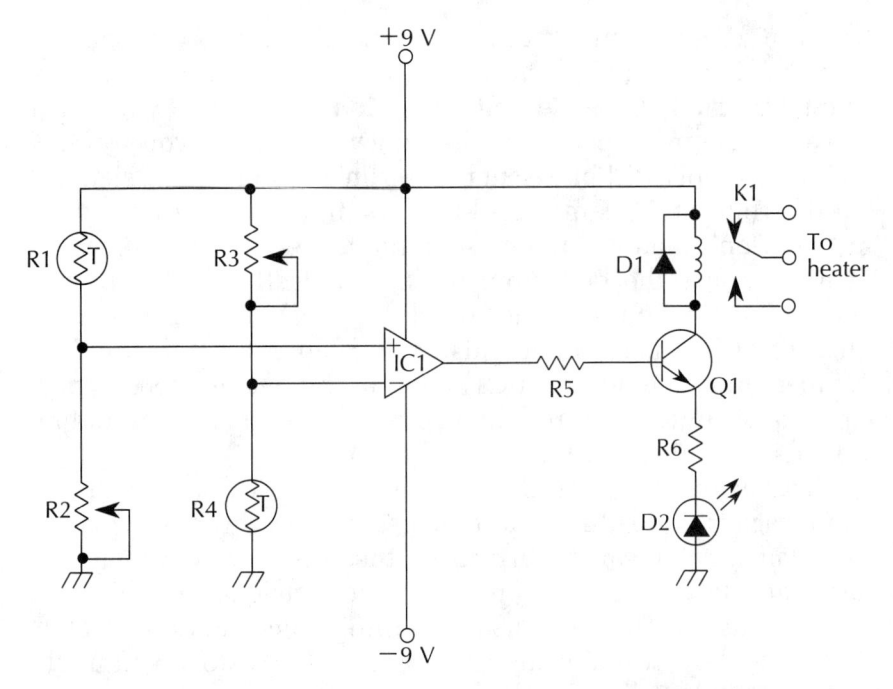

Fig. 3-11 Project 20. Water-heater controller.

Table 3-8 Suggested parts list for Project 20. Water-heater controller.

IC1	op amp
Q1	NPN transistor (2N3904 or similar)
D1	diode (1N4002 or similar)
D2	LED
K1	relay to suit load
R1, R4	thermistor
R2, R3	500 kΩ trimpot—see text
R5	10 kΩ ¼ W 5% resistor
R6	390 Ω ¼ W 5% resistor

741 will work every bit as well. This op amp is wired as a simple voltage comparator. It compares the voltage at the midpoint of two simple resistive voltage divider networks. One of these voltage dividers is made up of R1 and R2, and the other comprises R3 and R4. All four of these resistive units are variable. R1 and R4 are thermistors, which change their resistance with their temperature, and R2 and R3 are simple manual trimpots. These trimpots permit you to calibrate the system precisely. If

you can accept a little more potential inaccuracy, you can use two fixed resistors in place of the trimpots. Both resistors should probably have the same value in this case, but you might have to do some experimentation to get the results you want. A pair of trimpots will make such experimentation a lot easier.

Thermostat R1 is mounted indoors, in the same room as the heater. Don't mount it too close to the heater, however, because it might sense the heat from the heater itself, and the project won't function quite as accurately as it should. The second thermistor (R4) is mounted outside, to keep track of the outdoor temperature. The op amp (IC1) looks at the two voltages, which are proportional to the indoor and outdoor temperatures and responds according to which is the higher.

Transistor Q1 is a simple amplifier stage to protect the op amp stage from loading, and to boost the current level up a bit, so a larger relay can be driven by the controller circuitry. Almost any low to medium power NPN transistor should work well in this circuit. If you use something other than the 2N3904 (a very widely available and inexpensive transistor), you might find you get better results if you change the value of resistor R3 somewhat.

The relay (K1) should be selected so its switch contacts can safely handle the required load—the aquastat in the heater. With some very large heaters you might need to cascade two relays. This circuit drives the smaller relay directly, and this smaller relay controls a second, heftier relay.

As with most relay circuits, diode D1 is included to protect the relay's coil from possible damage due to back EMF. Almost any silicon diode will do the job just fine.

When the relay is activated, LED D2 will light up. Resistor R6 protects the LED by limiting the current flowing through it. If you don't want the visual indicator, you can eliminate D2 and R6 from the circuit, and simply connect the emitter of Q1 to ground—either directly or through a very low-value resistor. As with any electronic project that comes into contact with water, you should use battery power only. Never use an ac power source with this type of project. If the monitored liquid gets into the circuitry itself, it could cause a potentially dangerous short circuit. Why take a risk of a fire or a painful, or even fatal electrical shock? There is no need for ac power because batteries will operate this circuit for such a long time before they need replacement.

❖ 4
Atmosphere-related projects

One of the most important elements of the environment is the atmosphere—the air everyone breathes. You don't usually think too much about the air. You can't see it, smell it, or taste it (unless it is very badly polluted), and it is always there in your day-to-day life. You couldn't live without it. If you have ever been denied air for a little while (a minute or two is more than long enough), perhaps by being underwater or in a very tightly enclosed space, you certainly appreciated the atmosphere when you got it back. (If you didn't get it back, you certainly wouldn't be reading this or any other book.)

Most of the time, you just take the atmosphere for granted. It doesn't seem to be very interesting in itself. But the atmosphere is actually quite complex, with many facets to it, some of which make various electronic projects worthwhile.

The projects presented in this chapter consider three aspects of the atmosphere—*wind* (air in motion), *humidity* (moisture content of the air), and *ionization* (electrical charges in the atmosphere).

Wind-speed indicator

Often you will need to analyze the wind conditions. Wind is simply air in motion, so this unquestionably comes under the heading of *atmospheric*.

Wind can have a significant influence on the subjective experience of weather. If the temperature is about 48°F, and the humidity is about 50%, it might seem like a moderately cool, but fairly pleasant day. But that's assuming the wind is still, or blowing at a very low speed. But suppose everything remains the same, but 25 mph (miles per hour) winds start billowing down from the north. It's going to feel a lot colder and unpleasant.

There are other reasons to be interested in the wind speed. For example, suppose you want to erect a tall roof-top antenna. If you want it to stay in place, it will have to withstand the highest wind speeds normal to the area. By measuring the wind speed on a few very windy days, you can get some idea of just what forces your antenna will have to hold up against.

A very crude wind-speed indicator can be made from a small permanent-magnet motor. In this application, you do not apply any voltage to the motor.

Instead, the motor is mounted upright (its shaft pointing up) outside, in the wind path. Make a wind-cup assembly, as shown in Fig. 4-1. This is simply a pair of crossbars, with a small cup-shaped paddle on each of the four ends. All four wind cups must be aimed in the same circular section, so no matter which way the assembly is rotated, the cups will always be facing the same way. It is very important to use lightweight (but sturdy) materials in the construction of the wind-cup assembly. It must be easily movable by the wind.

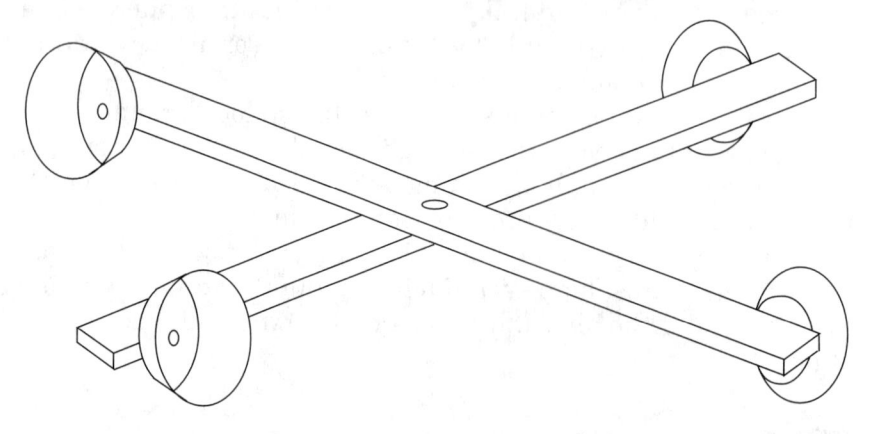

Fig. 4-1 *A simple wind-cup assembly can be made from a pair of crossbars with a small cup-shaped paddle on each of the four ends.*

This wind-cup assembly is mounted on the shaft of the motor, as shown in Fig. 4-2. Now, when wind blows past the motor, it catches one of the wind cups, and pushes it away, which brings the next wind cup into position to be pushed by the wind. In other words, the wind forces the wind-cup assembly to rotate, at a rate proportional to the wind speed. Obviously, the

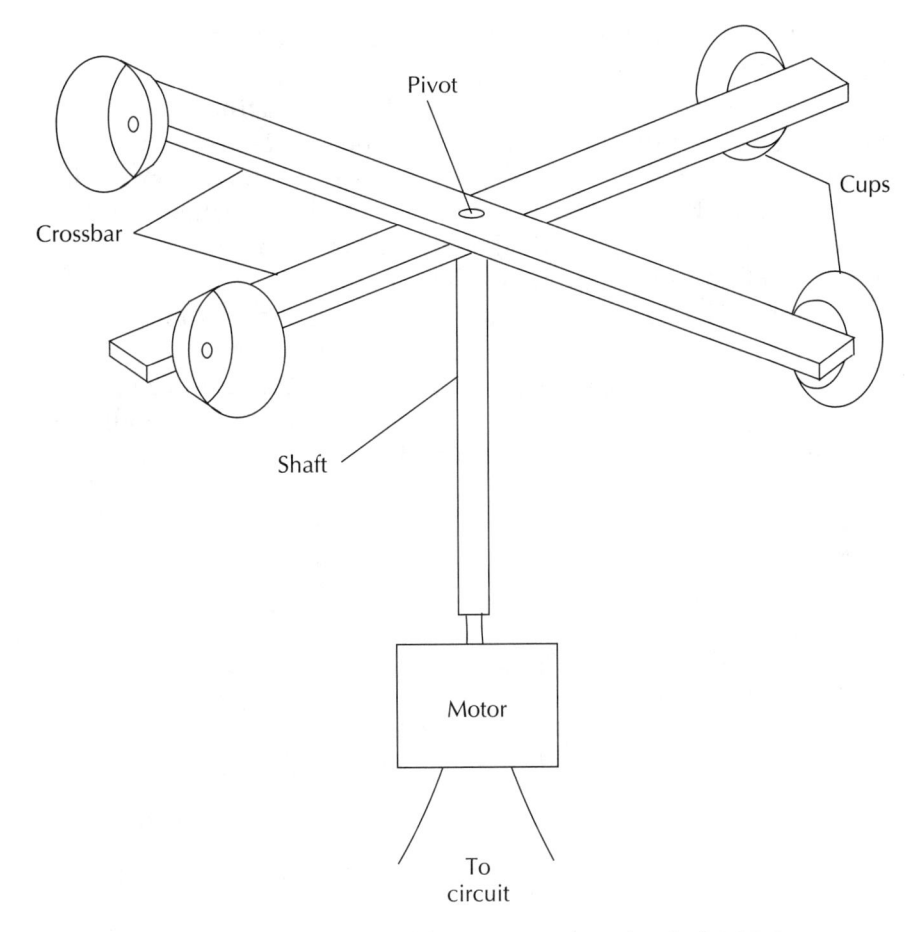

Fig. 4-2 *The wind-cup assembly is mounted on the shaft of the motor.*

faster the wind blows, the faster the wind-cup assembly will spin. Naturally, because the wind-cup assembly is attached to the shaft of the motor, the motor shaft must also rotate at the same rate.

The physical principles behind the operation of a permanent-magnet motor work both ways. Ordinarily, an external voltage is fed into the motor armature coil, causing the shaft to rotate. But if the shaft is mechanically rotated, a voltage will be induced in the armature coil. The faster the shaft is rotated, the higher this induced voltage will be. Now it is a simple matter to measure this voltage to get a reading proportional to the detected wind speed.

This will work, but it's rather crude and inexact. Study the simple wind-speed indicator in some detail, because you will be using some of the same principles in your actual wind-speed indicator project. Specifically, you will be using the same wind-cup assembly. But this time, you will not be using the permanent-magnet motor. Instead, the wind-cup assembly will be given its own shaft, at a 90-degree angle from the center of the unit. This shaft is passed through a grommet or wheel, with a hole large enough to permit the shaft to rotate freely, but small enough to hold the assembly upright.

Another shaft is mounted on the bottom end of the shaft, at a 90-degree angle. On one end of this bottom shaft mount a small permanent magnet. On the end, mount a counterweight, which balances the weight of the magnet. This new, modified wind-cup assembly is shown in Fig. 4-3.

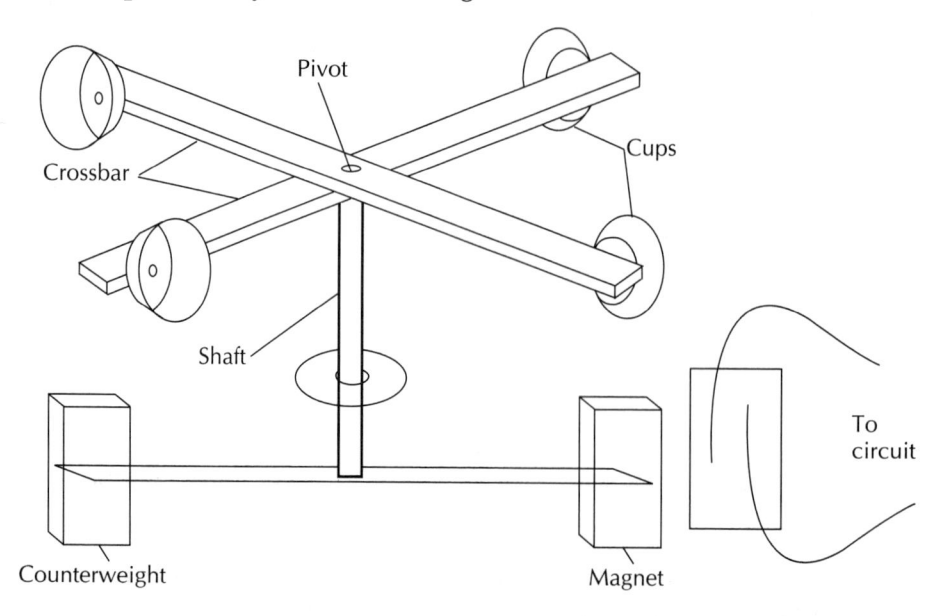

Fig. 4-3 *An alternate wind-cup assembly.*

Mechanically, this wind-cup assembly works in exactly the same way as in the previous version. As it rotates, from the force of the wind, the magnet is moved towards and away from the magnetic reed switch (S1). When the magnet is close enough to the reed switch, it will be activated, causing its contacts to close. When the magnet is too far away, the reed switch will be deactivated, and its contacts will be open. The magnetic reed switch should be physically positioned so that the magnet activates once per complete revolution of the wind-cup assembly.

Now all you need is a way to count how many times the

magnetic reed switch is closed during some consistent reference time. This timing will give you a relative indication of the wind speed. Counting the revolutions is easy enough to do electronically. All you need is a digital counter circuit, using the switch closures as the input pulses to be counted. You will also need a timer circuit (specifically, an astable multivibrator circuit) to establish the reference period. The counter circuit should be reset to zero at the end of each reference period, so a new counting cycle can begin. Finally, you need some way to display the count per cycle for the output of the entire circuit.

There are many possible ways to accomplish these functions electronically. There were a number of decisions to be faced in designing the actual circuitry for this project. Initially, a numerical read-out was envisioned for the project, using seven-segment LEDs or LCDs, but unfortunately this proved impractical. If you are going to use numerical outputs in a project like this, they should be calibrated in meaningful units. For a wind-speed indicator, the obvious choice would be to have the output read in miles per hour. But the calibration procedure seems far too complicated for the value of the project. A lot of expensive conversion circuitry would be required. The final project would be ridiculously expensive and would be difficult to calibrate.

Fortunately, you don't really need a full numerical read-out giving the exact wind speed. In most cases, you only need to know the approximate wind speed. In the final version of this project, a range of sixteen equal but arbitrary step-ranges are used, indicating in a pseudo bar graph fashion by a string of sixteen LEDs. On each timing cycle, the lowest LED will light up, then it will go out and the next LED will light up, going out as the next higher LED lights up, and so forth. The highest LED that lights up indicates the range of the current wind speed. When one timing cycle ends and the next begins, the display will suddenly drop from the highest LED (for the detected wind speed) down to the lowest LED again.

This display might seem a little odd and difficult to read at first, but most people can get the hang of it in a minute or so. A more stable display is possible—for example, one that lights up only the highest LED to indicate the present wind speed range—but this feature would add to the overall cost of the project, and there would be little real advantage gained.

Advanced and ambitious experimenters might want to try their hand at adapting this project for a stable display output, or even a calibrated numerical display. You might come up with another technique.

The schematic diagram for my version of the wind speed indicator circuit is shown in Fig. 4-4. A suitable parts list for this project is given in Table 4-1.

Ideally, all four of the ICs in this circuit should be of the CMOS type, as specified in the parts list. If CMOS devices are used, the circuit supply voltage can be anything from +9 to +15 V. However, you might have difficulty locating the 74C90 and the 74C154 chips. If you do have such problems, you can substitute the standard TTL versions of these devices—the 7490 and the 74154. In this case, the circuit supply voltage must be a well-regulated +5 V. TTL ICs have little or no tolerance for other supply voltages, and could be damaged or destroyed if the supply voltage isn't very close to +5 V.

The pin numbers for the CMOS and TTL versions of these chips are identical. No changes in the circuitry are required to use the TTL ICs in place of the CMOS devices listed in the parts list, except for the critical change in the supply voltage for the entire circuit.

IC1 through IC3 MUST be of the same type. That is, do not mix CMOS and TTL chips in the same circuit. It will not work (without modifications to the circuitry), and the different supply voltage requirements will result in unnecessary complications at best.

IC4 is specified as a 7555 timer. This is a CMOS version of the popular 555 timer IC. You can substitute a regular 555 chip for the 7555 suggested in the parts list. No changes in the circuitry are required to support such a substitution. The 7555 and the 555 timer ICs are pin-for-pin compatible. Either will work on either the CMOS or TTL supply voltages, so your choice of which chip to use for IC4 is not dependent on what you used for IC1 through IC3. If you use a 555 chip for IC4, capacitor C4 probably won't be necessary, but it certainly won't do any harm.

None of the passive components (resistor or capacitor) values are particularly critical in this circuit. Capacitors C1 through C4 are simply power supply-line filters. Their purpose is to filter out any possible transients that might appear riding on the supply voltage. Such transients can occur even with battery power due to rf interference. Such a noise spike can sometimes confuse a digital chip, causing it to operate incorrectly, usually just for a second, but such a glitch would still throw off the counter. Use of the filter capacitors is particularly important

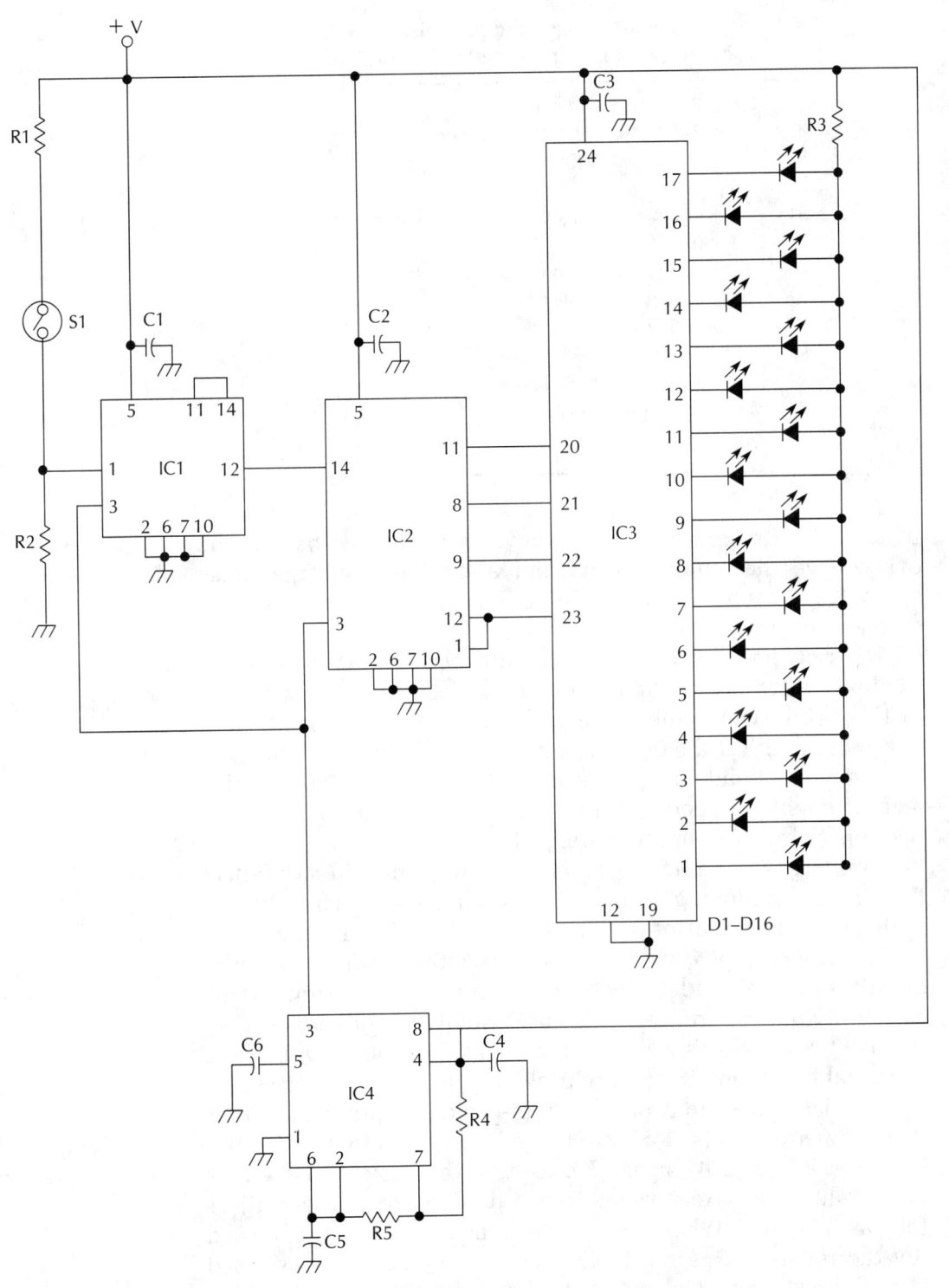

Fig. 4-4 Project 21. Wind-speed indicator.

Table 4-1 Suggested parts list for Project 21. Wind-speed indicator.

IC1, IC2	74C90 decade counter
IC3	74C154 4-line to 16-line decoder
IC4	7555 CMOS timer
D1–D16	LED
S1	Magnetic reed switch (normally open, SPST)
C1–C4	0.1 µF capacitor
C5	4.7 µF 25 V electrolytic capacitor
C6	0.01 µF capacitor
R1	100 Ω ¼ W 5% resistor
R2	2.2 MΩ ¼ W 5% resistor
R3	330 kΩ ¼ W 5% resistor
R4	3.3 kΩ ¼ W 5% resistor
R5	150 kΩ ¼ W 5% resistor

when TTL ICs are used. These devices cannot withstand the overvoltage that could result from a particularly large noise spike. One or more of the ICs in your circuit could be instantly damaged or destroyed.

Each of the filter capacitors should be mounted physically as close as possible to the power supply lead of the IC it is protecting. The exact value of these filter capacitors is irrelevant, and will not affect the operation of the circuit in any noticeable way. If you intend to use the project in an electrically noisy area, it might be a good idea to increase the value of these capacitors so they can provide more filtering.

Similarly, the value of capacitor C6 does not directly affect the circuit functioning. This is simply a bypass capacitor for the voltage control input of the 7555/555 timer. You are not using the voltage-control capabilities of this chip in this particular circuit. If this pin is left floating, the timer might exhibit some instability under certain conditions. Usually, this bypass capacitor isn't really necessary, but it is cheap insurance against the potential frustration if stability problems do happen to arise.

Normally, the input of the counter circuit (pin 1 of IC1) is held low through resistor R2. This is interpreted by the counter circuit as a logic LOW signal. Notice that this resistor has a very large value. The exact value is not particularly important, as long as it is large. When the magnetic reed switch (S1) is closed (by the revolving magnet in the wind-cup assembly described above), a HIGH pulse is applied to the input of the counter. This

HIGH pulse is derived from the circuit supply voltage through resistor R1. Again, the value of this resistor isn't very important, but it must be small, especially as compared to the value of resistor R2. In some cases, you might be able to eliminate resistor R1 from the circuit altogether, although eliminating it isn't really recommended.

Resistor R3 is a current-limiting resistor for the LEDs (D1 through D16). IC3 will only light up one LED at any given instant, so only a single current-limiting resistor is required. You might want to experiment with the value of this resistor a little. Increasing its value will reduce the brightness of the LEDs when lit. Similarly, to make the LEDs brighter, decrease the value of resistor R3. Certain restrictions apply here. Do not make the value of this resistor less than about 100 Ω, or it might not limit the current sufficiently. One or more of the LEDs could be damaged, or even destroyed. On the other hand, if R3 has a value above about 1 kΩ, the LEDs probably won't glow brightly enough to be visible, which would render the entire project quite useless.

This leaves just three passive components to consider in this circuit. These are resistors R4 and R5 and capacitor C5. These are the frequency determining components for the timer stage (IC4). You can read about the values of these timing components in this chapter.

IC1 is wired as a divide-by-ten counter. It produces one output pulse (at pin 12) for every 10 pulses it receives at its input (pin 1). This frequency division extends the upper range of this wind speed indicator project. If you are more interested in very low wind speeds, you should eliminate this IC from the circuit and apply the input pulses from the magnetic reed switch (S1) directly to pin 14 of IC1.

IC2 is wired as a four-bit binary counter. It counts the pulses at its input (pin 14) and counts from 0000 to 1111 at its four outputs (Pin 12, 9, 8, and 11).

You might have noticed that even though the same IC (the 74C90) is used for both IC1 and IC2, there are different pins identified for these two ICs. The pins are different because the ICs are being operated in different modes.

The four bit output from IC2 is fed to IC3, which is a 4-line to 16-line decoder. That is, it takes a four-bit binary input and activates one of sixteen possible outputs. For each possible combination of input bits (ranging from 0000 to 1111), the 74C154 brings one and only one of its outputs HIGH, while the other fifteen outputs are held LOW. Only the LED receiving the HIGH signal will be lit. The others will all be dark.

When a timing cycle begins, the count will, by definition, be 0000, so LED D1 (connected to pin 1 of IC3) is lit. LEDs D2 through D16 are dark. When the wind-cup assembly is rotated ten times, IC1 feeds one pulse to IC2, which counts it, and sends a value of 0001 to IC3. IC3 now extinguishes D1 and lights D2 (connected to pin #2). On the next pulse fed to IC2, the count goes to 0010 and only LED D3 (connected to pin 3) will be lit.

This general process will continue until the timing period is concluded. At this time, the highest LED in the sequence will be lit, indicating the approximate wind speed. The counters will then be reset to 0000, and only D1 will be lit again. The cycle will then repeat.

To read the output of this wind speed indicator project, just watch to see which LED is the highest that is lit up in the repeating sequence. This will tell you roughly how fast the wind is blowing.

The timing cycle in this circuit is controlled by IC4, which is a simple, common astable multivibrator circuit. It generates a very narrow width pulse wave. During most of each timing cycle, the output of IC4 (pin 3) is LOW. Once per timing cycle it goes HIGH for a very brief burst. This HIGH pulse is fed to pin 3 of IC1 and IC2 to reset the counters to 0000.

If this reset oscillator was not included in the circuit, all sixteen LEDs would always light up in a repeating sequence, regardless of the detected wind-speed. If the wind speed is very low, the sequence will occur rather slowly, and a high wind speed will cause the circuit to step through its output cycle very fast, but this is hardly a conveniently readable output. You might as well just look at the wind-cup assembly and try to guess how fast it is spinning.

The value of resistor R4 should be very small to keep the HIGH pulses as narrow as possible. Don't reduce the value of this resistor below 1 kΩ as a bare minimum. The timer circuit might not be able to function correctly with a smaller resistance.

The total length of the timing cycle from a 555 (7555) astable multivibrator circuit can be found with this formula (the component numbers are taken from your present project, of course):

$$T = 0.693 C5 (R_4 + 2R_5)$$

Using the component values suggested in the parts list you get a timing period approximately equal to:

$$T = 0.693 \times 0.0000047 \times (3300 + 2 \times 150{,}000)$$
$$= 0.693 \times 0.0000047 \times (3300 + 300{,}000)$$
$$= 0.693 \times 0.0000047 \times 303{,}300$$
$$= 0.99 \text{ second}$$

Allowing for component tolerances, you might as well round this off to 1 second. The frequency is therefore 1 Hz, of course:

$$F = \frac{1}{T}$$

The counters will be reset about once per second. A once-per-second update of the reading is a pretty good choice. You will be able to see the flashing LEDs clearly, but you won't have to wait a long time for the circuit to step through an entire cycle to reveal the highest lit LED in the sequence.

For many applications, no calibration is really needed. You can just say, "the wind speed is in range 3," or whatever, and you'll have a good idea of the relative wind speed. But if your application calls for a more absolute measurement, it would help to have some idea of just what kind of speeds range 3 involves.

In this project, calibration doesn't involve any circuit adjustments; it involves labeling the LED read-outs.

There are a number of ways you can calibrate an instrument of this type. The easiest way would be to compare it with a calibrated *anemometer* (wind-speed meter). Unfortunately, most people don't have access to a calibrated anemometer, and if you did, you probably wouldn't be very interested in this particular project in the first place.

Another approach is the automobile method. Warning—this method of calibrating the project absolutely REQUIRES two people. Do not attempt to do it by yourself. That could be very, very dangerous, no matter how good a driver you think you are. DON'T DO IT!

Pick a relatively windless day. The calmer the wind, the better. As one person drives the car along an open stretch of highway (preferably with as few turns as possible), the second person sits in the passenger seat and holds the wind-speed indicator project, so the wind-cup assembly is sticking out the window. As the car is driven at various constant speeds, the second person notes which is the highest LED lit up for each speed. Because the range steps are of equal size, you only need to determine two or three of them, and you can interpolate the rest. Then its just a matter of marking the appropriate speed value under the appropriate LED on the project case. For example, if each LED represents a 6 mph range, the LEDs would be marked something like this:

D1	(pin 1)	0 mph
D2	(pin 2)	6 mph
D3	(pin 3)	12 mph
D4	(pin 4)	24 mph
D5	(pin 5)	30 mph
D6	(pin 6)	36 mph
D7	(pin 7)	42 mph
D8	(pin 8)	48 mph
D9	(pin 9)	54 mph
D10	(pin 10)	60 mph
D11	(pin 11)	66 mph
D12	(pin 13)	72 mph
D13	(pin 14)	78 mph
D14	(pin 15)	84 mph
D15	(pin 16)	90 mph
D16	(pin 17)	96 mph

The step range you're likely to come up with in your project depends on many factors, including the various dimensions of your wind-cup assembly, and the exact component values (with the tolerances factored in) of the frequency determining components in the timer circuit.

Some hobbyists might not be satisfied with whatever step units they happen to get, which could be quite awkward. You can make the circuit such that it can be more precisely calibrated by making the timing cycle manually variable. Use a fixed-value resistor of about 100 kΩ in series with a 100 kΩ to 250 kΩ trimpot in place of resistor R5. Varying the setting of the trimpot will adjust the timing cycle period, which will affect how many count units (how many LEDs) will get through per cycle for a given wind speed. For example, you could adjust the timer frequency so each step range (each lit LED in the sequence) equals a wind speed of 5 mph or 10 mph—that's certainly a lot neater.

Once again, if you are primarily interested in very low wind speeds only, simply omit IC1 (and capacitor C1, of course) from the circuit entirely. This eliminates the divide-by-ten input scaling function.

What happens if the wind speed exceeds the maximum range of the instrument? For example, using the 6 mph per step range example, suppose the monitored wind speed was 115 mph (a pretty outrageous wind speed for most natural environments). The LEDs would cycle on and off from D1 through D16, then on the next count, the counters would automatically cycle back to 0000, lighting up D1 again, (indicating a wind speed of 104 mph). Then D2 would light up (wind speed is 110 mph), and finally LED D3 would light up as the maximum range, (116 mph), before the counters are forcibly reset by IC4, and the entire sequence starts over. This would be very confusing to read, although possible. Fortunately, such an occurrence is quite unlikely (unless you've deleted IC1 from the circuit). How often are you likely to try measuring wind speeds above 100 mph? Not many people live in wind tunnels. It is doubtful that a homemade wind-cup assembly would even hold up to such high wind speeds anyway. And if the wind speed is sufficient to destroy your wind-cup assembly, you can't measure it even if the circuit range is extended far enough.

In most practical uses, you'll probably find that LEDs D11 through D17 seldom if ever light up.

If you eliminate IC1 from the circuit to make it more sensitive to low wind speeds, you are much more likely to run into overrange conditions, but they won't do any harm—it will just be difficult (in some cases) impossible to get a meaningful reading until the wind speed drops back into the acceptable range for your project.

An imaginative electronics hobbyist could come up with a lot of very interesting and unusual applications for the basic principles used in this project.

Humidity meter

Usually, when you think about comfortable or uncomfortable weather, you mostly think about the temperature. Temperature certainly is important in determining your comfort level, but not in absolute terms. Under some conditions 40°F can feel a lot colder than 30°F, and 85°F can be unbearably hot, and 95°F doesn't seem nearly as bad. Obviously it is not the temperature that determines your comfort level. What is the deciding factor? Humidity.

Relative humidity is a measurement of how much moisture the air contains. It is normally given as a percentage, with 0% humidity representing perfectly dry, moisture free air, and 100% signifying the maximum amount of moisture the air can possibly contain. Relative humidity is used, rather than absolute humidity, because the percentages can vary quite a bit depending on a number of external factors. For example, temperature and barometric pressure can make a big difference in how much moisture the air can hold at its 100% point.

In hot weather, high humidity can make the subjective temperature feel hotter than it actually is. This is because high humidity neutralizes the effects of your perspiration. Humans are designed so that when you get hot, you start to sweat, as an automatic self-cooling procedure. The sweat coats your skin with moisture. As this moisture evaporates, the skin will be cooled a little. But if the relative humidity is high, the surrounding air will be less capable of absorbing the moisture in the evaporation process. If the relative humidity is 100%, there will be no evaporation at all, because there is no place for the moisture to go. It just stays on your skin. Not only do you not get the cooling evaporation effect, but the accumulated sweat makes you feel clammy and even hotter and more uncomfortable.

With cold weather, the situation is a little more complex. Popular folk wisdom usually claims that low humidity makes cold temperatures feel colder, but this is only partially true, under some circumstances. Very low humidity at cold temperatures will make the subjective temperature seem crisper and more biting. However, very high humidity can make cold air feel colder. High humidity means that there is a lot of moisture in the air. The air is damp. Have you ever had somebody pour water on you while you're cold? It makes you feel a lot colder, doesn't it? Everyone would probably agree that being cold and wet is more miserable than being cold and dry. Very humid air is just a milder version of being wet on a cold day.

Most people will find that cold air feels the least cold when the relative humidity is close to the 50% point—not too much moisture in the air and not too little.

Very low humidity tends to make problems with static electricity more severe. Many common forms of artificial heating for a room or building will tend to dry the enclosed air. That is, the relative humidity will be reduced by the heating process.

As you can see, humidity is an important factor of the environment. Often it would be helpful to have some way to measure the relative humidity. Obviously, that is the function of the next project.

Remember, although this device is called a "humidity meter," it measures relative humidity, not absolute humidity. Actually, measuring the absolute humidity would really be useful in practical applications. Relative humidity is what counts in determining the comfort level, and the degree of static electricity build-up. The maximum amount of moisture the air can hold varies with the temperature of the air. For example, 25% humidity at one temperature would be 48% humidity at a different temperature, even though the absolute amount of moisture in the air remains exactly the same.

Humidity meters are sometimes called *hygrometers*, which is actually a bit more accurate a description, but it is a more obscure and less understood term for the average person, so *humidity meter* is probably a better choice. It gets the concept of the device across more conveniently.

The tricky part of designing an electronic humidity meter is devising a suitable sensor device. Electronic components are not normally responsive to changes in relative humidity. Many are sensitive to moisture—if they get wet, they will be damaged, or the component value will change excessively. Water can cause a short circuit, because it has a fairly low resistance. For this reason, many electronic components are packed in more or less water-tight housings. But if you don't use the water-tight housing, that doesn't help you in developing a humidity sensor. It is not like semiconductors and light. Normally semiconductors are enclosed in a light-tight housing to prevent light sensitive effects, but in a photosensor, a lens in the housing deliberately lets light in. For moisture sensitivity, there are just two possible states—the component works properly when it's dry, or it fails to work when the moisture level exceeds some specific point. This type of yes/no response would be of no practical use in a meter circuit. In addition, the odds are the sensor will be permanently damaged the first time the moisture level exceeds its limits. In other words, the moisture detector would only work once before you need to replace the sensor. This is hardly an efficient or intelligent approach to designing a useful humidity meter.

Obviously, you can't use any existing standard electronic components to come up with a practical humidity sensor. You need a device specifically designed for the purpose. A few humidity sensors have been made commercially available, but they tend to be difficult for the hobbyist to find, and they also are quite expensive.

Fortunately, it is not too hard to build a simple homemade humidity sensor. It might be a bit crude, and not really very precise, but it does the job for most general purposes. The difference between 73% relative humidity and 77% relative humidity is rather slight, and essentially negligible for most practical purposes. For example, you'd almost certainly notice no difference at all in the comfort level by such a change in the relative humidity in your immediate environment.

Will you have trouble finding the necessary parts to build your humidity sensor? Probably not. The main active ingredient is salt—that's right, plain, ordinary table salt (sodium chloride). Everything else is just mechanical, and can easily be redesigned to suit the materials you happen to have available.

Most salts (including good old table salt) are naturally *hygroscopic*, which means they tend to absorb or emit moisture into the surrounding air to equalize the vapor pressure. That is, the salt will try to make its relative humidity match that of the surrounding air. This might not seem like too much help for your purposes here. You've just moved the relative humidity from the air to the salt—you still need a way to measure it electronically.

Fortunately, this is very easy to do. Like most substances, salt has a resistance, which can be read across it. As the salt absorbs moisture, its resistance will drop proportionally, in a predictable and reasonably linear manner.

There are many ways you can electronically measure the resistance of the salt. In this project, you will use a simple bridge circuit. But before you get to the circuitry, finish building your humidity sensor.

A simple type of construction you can use for a practical humidity sensor is shown in Fig. 4-5. The base of the sensor can be made from a piece of circuit board, measuring about 2 inches by ½ inch. A good, convenient thickness would be ¹⁄₁₆ inch, which is pretty standard. The circuit board should not be copper clad, of course, because the copper would create an electrical short circuit, so the measured resistance would also be 0 Ω (or very close to it), regardless of the relative humidity. Certainly there wouldn't be any point at all to that.

In the following measurements, you will be assuming the base board is exactly 2 inches long. If you are using a longer board as the base, you will need to accommodate the outer dimensions accordingly.

Drill two mounting holes in the baseboard ⅛ inch in from either side. Two more holes need to be drilled into the base, ⅜ inch in from each mounting hole (or ½ inch from the closest end of the board). These holes accommodate a pair of 2-56 × ⅜ inch bolts, as shown in the diagram. The bolts should be made of

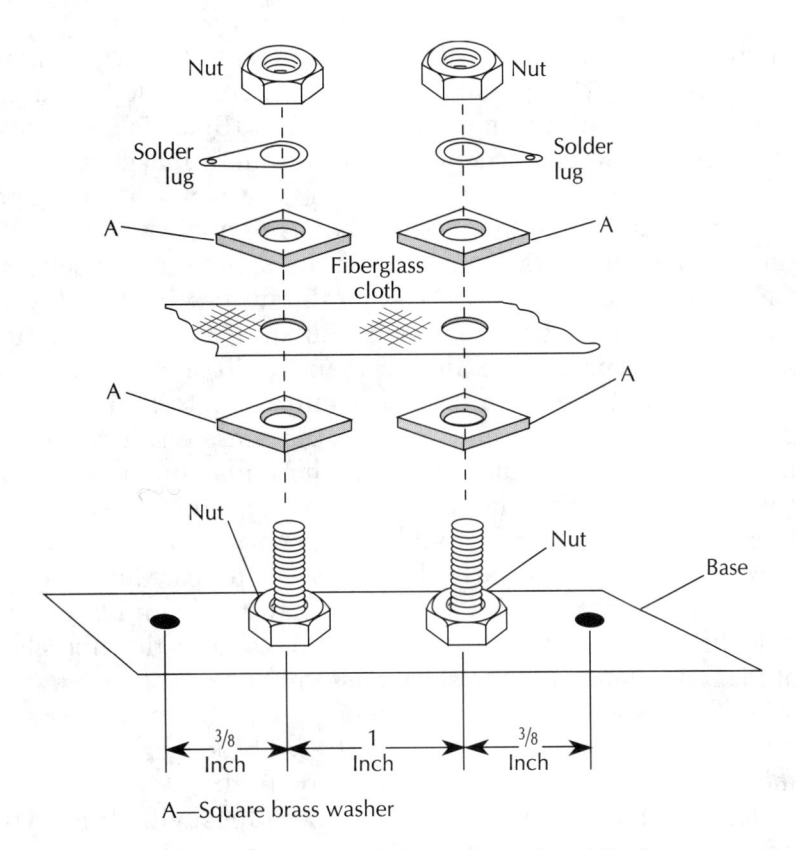

Fig. 4-5 *A relatively simple homemade humidity sensor you can build yourself.*

brass or stainless steel. A #44 drill would be suitable for drilling all four holes. If you follow the dimensions given here correctly, the bolts should be equidistant from the center of the baseboard, and exactly one inch from each other. This is the most critical dimension in constructing the humidity sensor. Do not add any variation here. It's not critical in the sense that you need absolute precision down to the millimeter level, but a 1-inch spacing is the measurement you want to aim for, as closely as practical, regardless of the length of your baseboard.

Hold each bolt in place on the board with an appropriate nut. The head of the bolt should be snug against the underside of the board. You might want to include a washer between the head of the bolt and the baseboard, but this usually won't be necessary.

Over the nuts tightened to the base, place a square brass washer on each bolt. These washers should measure as close to ½-inch square as possible, and should have a thickness of

1/32 inch. It might be a little difficult to find square brass washers, but it shouldn't be impossible. In a pinch, you could use larger strips of brass, and cut them down to size. (Brass is fairly easy to cut.) Or, if you can accept just a little more inaccuracy, you could use round washers if you absolutely have to. The most important thing is that the washers be made of brass. You might be able to substitute stainless steel washers, but brass is better.

Next take a piece of fiberglass cloth cut to ½ inch wide and 1¼ inch long. It might be easier if you start out with a ½-inch wide strip that is a little longer than you'll need. The cloth should be about the same thickness and weight of an ordinary handkerchief. Place the cloth, above the brass washers, across the bolts as shown in the diagram. The bolts should protrude through the cloth. You should be able to force the bolts through the cloth, but if you have trouble, you can use an awl or other sharp object to make a starter hole. Try not to tear the cloth excessively or make too large a hole. A small tear shouldn't make much difference, but try to avoid it. Make sure the fiberglass cloth is lying flat across the sensor assembly.

The next stage is to add another pair of brass washers on each bolt, over the top of the cloth strip. These washers are the same as the earlier pair you have already used.

Place a solder lug over each nut. These lugs will permit you to make the necessary electrical connections, of course.

Finally, complete each bolt with an appropriate nut, tightened sufficiently to hold the entire assembly securely in place. Trim off any of the fiberglass cloth extending out from the edges of the washer, but not the bridge extending between the two washers. Remove as many weave threads from the cloth bridge between the washers as you can. This will permit the sensor to respond to changes in the relative humidity faster.

The next step is to soak your sensor assembly in a solution of lithium chloride. If you are wondering where the heck you're going to get that exotic-sounding chemical, don't worry about it. Just take a mound of salt about the size of a dime and dissolve it in a tablespoon of water. *Voilà!* You've got a solution of lithium chloride.

Soak the sensor, then shake off any excess solution, and let it dry thoroughly while you construct the rest of the project. The fiberglass cloth now has the salt embedded in it, and you can measure the resistance between the two solder lugs.

If you are a little confused about the construction of the humidity sensor, just carefully reread the instructions a couple times, then work step by step. This is one of those cases where describing the construction procedure is more complicated than actually doing it. It is really not that difficult at all.

Now, look at the actual circuitry for your humidity-meter project. The schematic diagram for this project is shown in Fig. 4-6. A suitable parts list for this project is given in Table 4-2.

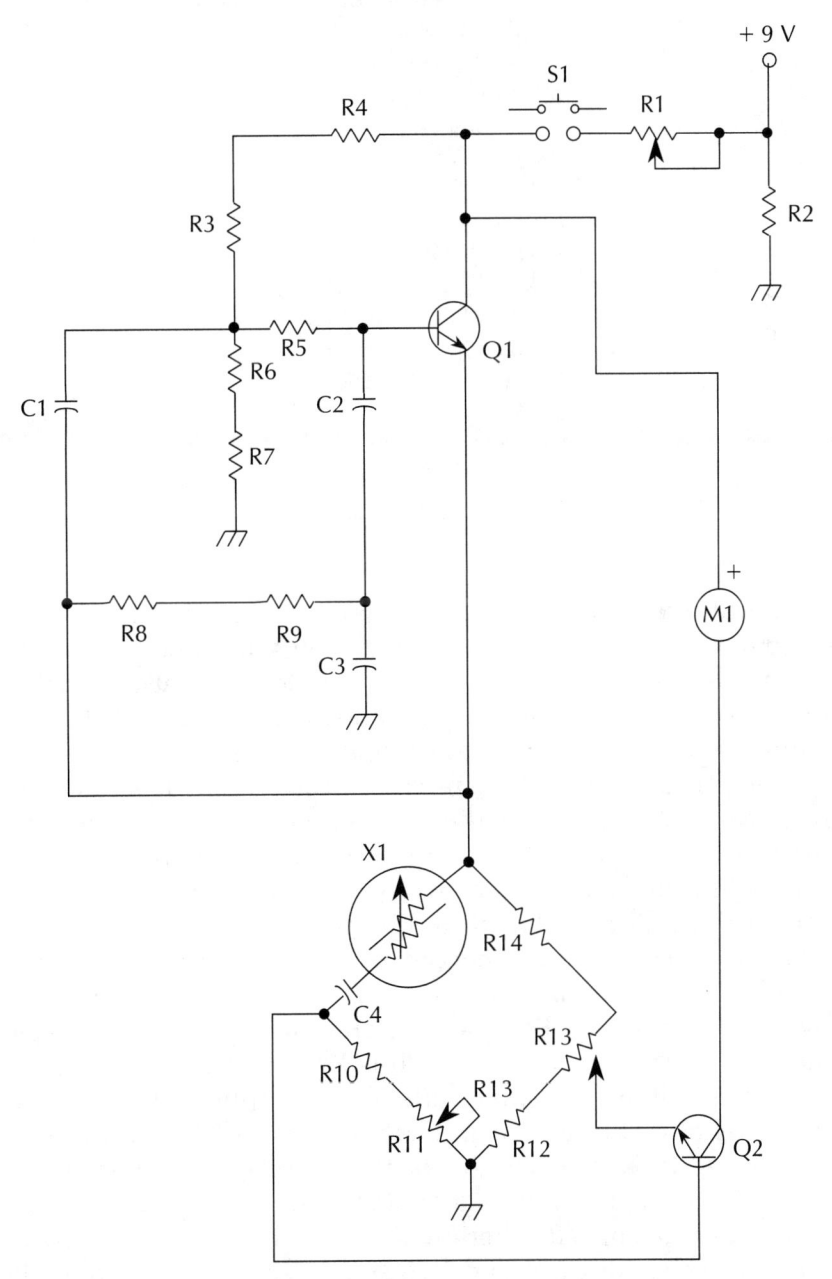

Fig. 4-6 *Project 22. Humidity meter.*

Table 4-2 Suggested parts list for Project 22. Humidity meter.

Q1, Q2	NPN transistor (2N3711 or similar)
M1	1 mA meter
X1	sensor—see text and Fig. 4-5
S1	normally open SPST push-button switch
C1, C4	0.1 µF capacitor
C2	0.0047 µF capacitor
C3	0.047 µF capacitor
R1	10 kΩ trimpot
R2	100 kΩ ¼ W 5% resistor
R3	15 kΩ ¼ W 5% resistor
R4	1 kΩ ¼ W 5% resistor
R5	51 kΩ ¼ W 5% resistor
R6	33 kΩ ¼ W 5% resistor
R7	3.3 kΩ ¼ W 5% resistor
R8, R10	2.2 kΩ ¼ W 5% resistor
R9, R12	220 Ω ¼ W 5% resistor
R11	250 kΩ trimpot
R13	1 kΩ trimpot
R14	3.9 kΩ ¼ W 5% resistor

Substituting any alternate component values in this circuit is not recommended because substitution will affect the accuracy of the completed instrument. This requirement especially applies for capacitor C4 and resistors R10 through R14. Use components with the lowest tolerance ratings as possible here.

You can substitute other similar transistors for Q1 and Q2, if you need to. The exact specifications of these transistors are not too critical, and any differences in a reasonably close substitution should be compensated for by the calibration procedure, which will be discussed later in this section.

Notice the odd symbol marked *X1* in the schematic diagram. Don't be thrown by this. It is simply the humidity sensor you have just constructed. Because it is an unusual device, it needs a nonstandard schematic symbol. It is X1 just to give it a name.

This circuit might seem a little more complicated than it really needs to be for measuring a dc resistance. The reason it is complicated is that you aren't exactly measuring a dc resistance in this project. Salts tend to act a little oddly to the continuous polarity of a dc current. There will tend to be a continual resistance change as long as the dc current is continuously applied.

Obviously, you wouldn't get any meaningful readings from your meter circuit that way.

Therefore, it is necessary to use an ac current for the test signal through the Wheatstone bridge meter circuit. But the project can be powered from a regular 9 V transistor radio battery. The necessary ac test signal is generated by an oscillator circuit made up of transistor Q1 and its associated components.

Using the component values suggested in the parts list, the oscillation frequency will be about 400 Hz. This isn't too critical, as long as the signal frequency is constant. If the actual frequency is considerably different from the 400 Hz design standard, you might have some difficulty in calibrating the circuit.

Notice that even though you are using an ac test signal, you are still measuring just the dc resistance. You don't have to worry about the inherent complexities of ac impedance in this application.

Transistor Q2 serves as a simple amplifier to boost the signal fed to the meter.

The circuit will produce a reading on the meter only when switch S1 is depressed. This helps extend the life of the battery, but isn't really essential. You can replace the push-button switch with a standard SPST on/off switch, if you prefer.

There are three trimpots in this circuit used to calibrate the circuit. Actually one of them is pretty much optional.

Adjust trimpot R1 so that the supply voltage passing through switch S1 is exactly 8.4 V. If you can accept a little slight inaccuracy, which might not even be noticeable, you can replace this trimpot with a 6.8 kΩ fixed resistor.

The humidity sensor you built earlier is one leg of a simple Wheatstone bridge circuit, along with resistors R10 through R14. Capacitor C4 compensates for the ac test signal explained above. When the sensor changes its resistance due to changes in the sensed humidity, the bridge will be thrown out of balance. This change will be sensed, buffered, amplified slightly by transistor Q2, and fed to the meter (M1) for display. Use a milliammeter with a full-scale value of 0.1 mA in this project. In this application, the meter will read relative humidity values ranging from 0% to 100%. You can make up a special dial plate for a direct read-out of percent, but this isn't completely necessary in this project, if you use a 1 mA meter. Reading the correct value in percent from the meter is just a simple meter of mentally shifting the position of the decimal point. A reading of 1 mA means 100% relative humidity, a reading of 0.5 mA indicates a relative humidity of 50%, a relative humidity of 85% would be displayed as 0.85 mA, and so forth.

Of course, to get meaningful readings from your humidity meter project, you must calibrate the Wheatstone bridge, by adjusting trimpots R11 and R13.

Connect the humidity sensor you constructed earlier to the rest of the circuit through a pair of wires about 7 to 10 inches long. If you use longer connection wires, you might run into interference problems. To minimize such problems, use a twisted pair or a shielded cable to connect the sensor to the meter circuitry.

To calibrate the unit, place the sensor inside a jar containing a piece of wet tissue. The sensor leads can come out under the lid of the jar. It doesn't have to be sealed 100% air tight, but it should be sealed as much as possible, without damaging the sensor lead wires. The tissue should be very wet—just soak it in water. The sensor unit should be close to, but not quite touching the wet tissue. Wait a few minutes to allow the sensor time to adjust to the very humid environment within the jar. Press push-button switch S1 to take a reading, and carefully adjust trimpot R13 for a full scale reading on the meter (100%, or 1.0 mA). Wait a few more minutes, and take another reading. You should get the same full scale reading. If not, readjust trimpot R13 accordingly, then wait a few more minutes before taking a reading. Repeat this process until you no longer have to readjust the setting of trimpot R13 anymore. The variation was due to the fact that the sensor had not fully stabilized for the earlier readings.

Now, remove the sensor from the jar and dry it out by placing it in front of a fan for several minutes. While you are waiting for the sensor to dry out, determine the local ambient humidity, either with another calibrated humidity meter, if available, or you can use a conventional bulb thermometer. In this case, first make a note of the ambient temperature. It should be somewhere close to normal room temperature.

Next wrap some water-soaked tissue around the bulb and place it in front of the fan, to be cooled by the moving air. Make a note of the lower temperature now indicated on the thermometer.

Compare the two temperature readings with Table 4-3 to determine the ambient humidity level. To save space in the table, there are only three dry bulb temperatures indicated near standard room temperature. Use whichever one is closest to your dry bulb reading, or you can interpolate between values.

Table 4-3 Comparison table for calibrating the humidity meter project as described in the text.

Bulb temperature (dry) °F	Bulb temperature (wet) °F	Percent humidity
65	44	10%
65	47	20%
65	49	30%
65	52	40%
65	54	50%
65	57	60%
65	59	70%
65	61	80%
65	63	90%
70	47	10%
70	50	20%
70	53	30%
70	56	40%
70	58	50%
70	61	60%
70	63	70%
70	65	80%
70	68	90%
75	50	10%
75	53	20%
75	56	30%
75	60	40%
75	62	50%
75	65	60%
75	68	70%
75	70	80%
75	72	90%

Obviously, this method is only going to give you an approximate value for the ambient humidity level, but it should be close enough for most practical purposes. It is about as good as you are going to do without access to some rather expensive meteorological equipment. Fortunately, exact accuracy will rarely be essential for most practical relative humidity readings, as long as it is reasonably close.

By this time, the sensor should be dry. To make absolutely sure, you might want to wait another half hour or so before continuing with the calibration procedure.

Once the sensor is satisfactorily stabilized to the relative humidity of the environment, depress switch S1, and adjust trimpot R11 so the meter shows the same humidity value you just got from Table 4-3.

Heater humidifier

In winter months, heaters are naturally put into heavy use, especially in the more northern parts of the country. (In Phoenix, Arizona, where I live, heaters are not as heavily used.) Of course, using a heater a lot inevitably drives utility bills up, and, perhaps less obviously, it also tends to drive indoor humidity down. Obviously, higher utility bills are something of concern to everyone, even aside from the broader ecological concerns of excessive energy consumption. But why should you care about decreased humidity?

If the effect is slight, it doesn't really matter much. A lot depends on what the original humidity levels were (before the heater was used), and what type of heating is used. Steam heat, for example, does not decrease the humidity nearly as much as electric heat, but electric heaters tend to be considerably more energy efficient than steam heaters. As so often happens in life, you must make some sort of trade-off here.

Excessive decreases in humidity levels can be quite significant and problematic. For one thing, dry air (very low humidity) is a poor electrical conductor compared to moist air (high humidity), which means a greater build-up of static electricity and more frequent static discharges, or shocks, and an increased tendency for some small objects (particularly certain types of fabrics) to cling to one another.

The human body is also humidity sensitive. The most healthy environment for most people has humidity levels in the 40% to 60% range. In the winter months (assuming a cold climate), the house is more or less sealed off, so there is limited air exchange between the indoors and outdoors environments. Heating units tend to dry up the humidity in the air, which is not replaced very quickly with more humid air from outdoors. Often the indoor humidity in winter can drop to 10% to 20%, which is extremely dry. This can irritate some sensitive membranes in the body, leading to increased incidents of sore throats, and other such ailments. In addition, the body's natural humidity sensitivity can make dry air (especially in the 50 to 70°F range) seem even colder than normal. This means people will probably tend to set the thermostat to a higher setting to feel comfortable. The heater works more (and consumes more energy), and of course, it dries up the air even more.

Raising the indoor humidity to about 50% will permit a lower thermostat setting for a given degree of comfort. This will save fuel, and will be healthier for anyone in the house.

However, contrary to what some people believe, increasing the humidity much above 50% will not increase the benefits. Actually, high humidity is as bad as, if not worse than, low humidity. Some people, having heard about the problems of winter low humidity, add so much artificial humidity that their homes become positively dank. They are doing themselves more harm than good. Very moist air (very high humidity) will feel colder for any given temperature. It will also tend to hold more germs and bacteria than drier air, which obviously is unhealthy. The ideal is moderate humidity—around 50%, give or take a little.

This project is designed to automatically add suitable amounts of humidity to heated air. The schematic diagram for your heater humidifier circuit is shown in Fig. 4-7. A suitable parts list for this project is given in Table 4-4.

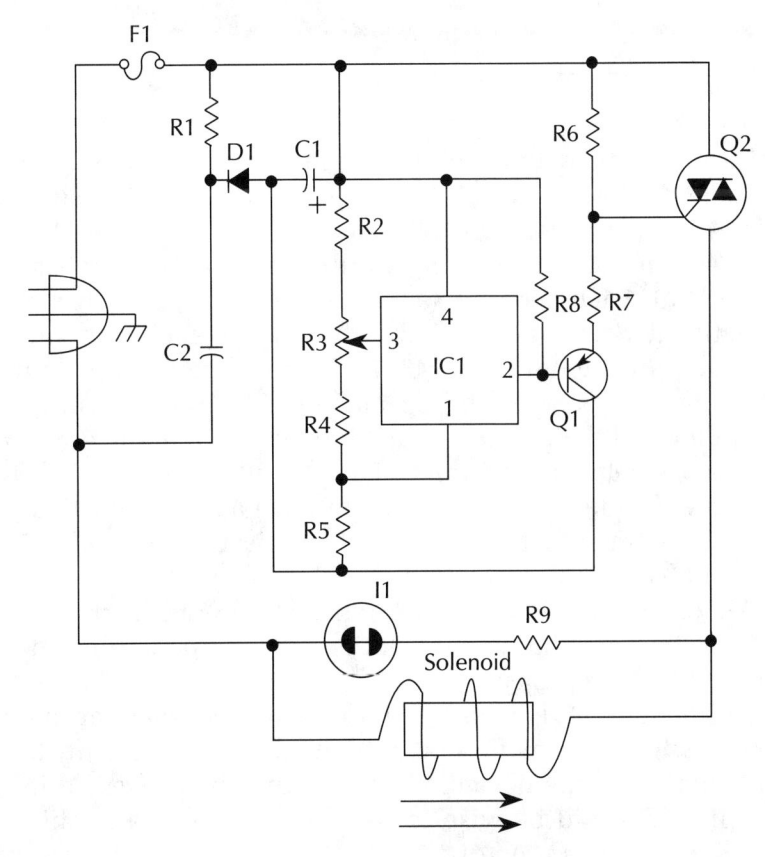

Fig. 4-7 *Project 23. Heater humidifier.*

Table 4-4 Suggested parts list for Project 23. Heater humidifier.

IC1	LM3911 temperature-sensor control IC
Q1	PNP transistor (2N3906 or similar)
Q2	triac to suit load (T2303 B or similar)
D1	diode (1N2069 or similar)
I1	NE-2H neon lamp
F1	3 A fuse
Solenoid	see text
Spray nozzle	see text (not shown in schematic diagram)
C1	100 µF 25 V electrolytic capacitor
C2	1 µF 200 V capacitor (non polarized)
R1	270 Ω ¼ W 5% resistor
R2, R4	27 kΩ ¼ W 5% resistor
R3	10 kΩ potentiometer
R5	4.7 kΩ ¼ W 5% resistor
R6	1.2 kΩ ¼ W 5% resistor
R7	1.8 kΩ ¼ W 5% resistor
R8	150 kΩ ¼ W 5% resistor
R9	27 kΩ ½ W 5% resistor

Most of the components in this circuit are common, readily available devices, and not too critical, so reasonable substitutions are acceptable. The one specialized component here is IC1, which is used as the temperature sensor for the project. IC1 is an LM3911 temperature-sensor control chip, and it is made by National Semiconductor specifically for such temperature sensing applications. The LM3911 is fairly well available now. Unfortunately, as with any specialized electronic component, the author and the publisher have no way to guarantee it won't become obsolete without warning in the future. This can dry up sources very quickly. Make sure you have a source for this part before investing any money into this project, or you might end up frustrated and disappointed.

Almost any low-power PNP transistor should work just fine for Q1. The operating parameters of this component are not at all critical in this project.

Component Q2 is a triac. Almost any triac that can handle the necessary current flow should work in this circuit, but if you substitute some device other than the one suggested in the parts list, it would be wise to first breadboard the circuit and make sure it works properly.

The passive components values (the resistors and capacitors) are also not too critical, so if you can't find a particular value, you should be able to substitute an appropriate series or parallel combination to come reasonably close to the value specified in the parts list. Actually, these are all fairly standard component values, and you shouldn't have much trouble finding any of them. Take note of capacitor C2, however. Even though it should be rated for 1 µF at 200 working volts (or more), you should not use a standard electrolytic capacitor here. Ordinary electrolytic capacitors are polarized devices (that is, with a definite positive lead and negative lead), and this component MUST be a nonpolarized unit, as the circuit carries current in both directions through this capacitor. Use a special nonpolarized capacitor, even though it will probably be a little more expensive and harder to find. In a pinch, you could use two 0.5 µF (or 0.47 µF) capacitors in parallel. Remember, capacitances in parallel add. These capacitor values might be a little easier to locate in some areas.

As always, do not omit or increase the rating of the fuse in this circuit. That could be very dangerous. As with any circuit carrying ac power, take all necessary precautions. Make sure everything is adequately shielded and insulated. No one should be able to touch any circuit conductor (including those nominally at ground potential) at any time while the project is in operation.

In using this project, the sensor (IC1) should be placed in the direct air path of the heater main vent (or vents). When the sensed temperature exceeds a point set by potentiometer R3, a solenoid is activated, opening a spray nozzle. Because the water is sprayed directly into the hot heating vent, it should vaporize almost immediately, adding to the air humidity in the heated room.

The spray nozzle should spray a relatively fine mist of water, for easy and quick vaporization. Otherwise, you'll just get a wet heating vent. If all the water isn't vaporizing, either you are spraying too much, or potentiometer R3 is set for too low a trigger temperature.

Suitable spray nozzles for this project can be purchased from almost any reasonably large hardware or plumbing supply store. They are usually fairly inexpensive. The solenoid valve can be cannibalized from an old washing machine (or something similar). Of course, you can buy a new replacement solenoid valve designed for such machines, if you prefer. Check with a local appliance repair person in your area, if you have trouble locating this part on your own.

Potentiometer R3 should be adjusted experimentally for the most comfortable overall humidity level. Be careful not to set this control either too high or too low, or you will defeat the entire purpose of this project, which is to create the most comfortable, energy efficient, and healthy indoor environment possible.

Air ionizer

You might have heard something about the effects of *air ionization* from time to time. Before you get to your air ionizer project, you need to take a little time to understand just what air ionization means.

First, what are ions? As you probably already know, all matter is made up of tiny particles, known as *atoms*. Atoms, in turn, are made of still smaller units, the major ones being *electrons*, *protons*, and *neutrons*. Each electron has a small negative electrical charge, and each proton has a similar small positive electrical charge. Neutrons are electrically neutral. The protons and neutrons of an atom are bunched together into a central cluster known as the *nucleus*. The electrons orbit around the nucleus, somewhat like planets around the Sun.

(This description isn't a completely accurate portrait of modern atomic theory, but it gets the basic concepts across well enough for your purposes here.)

Ordinarily, the overall electrical charge of an atom is neutral; that is, the total negative charges (from electrons) within the atom are exactly equal to the total positive charges (from protons). In other words, there are exactly as many orbiting electrons as there are protons in the nucleus. The positive and negative electrical charges completely cancel each other out, leaving a net electrical charge of zero.

But under some conditions, an atom might lose one (or possibly more) of its orbiting electrons. In this case, the atom obviously has more protons than electrons, so the total internal positive charges outweigh the total internal negative charges. The atom as a whole has a positive electrical charge. Such an atom is called a *positive ion*.

Similarly, it is sometimes possible for certain atoms to pick up one (or possibly more than one) extra electron, perhaps by stealing it from another nearby atom. With the extra electron (or electrons), the atom now has more electrons than protons, so the internal negative charges are greater than the total internal positive charges. The atom as a whole has a small negative electrical charge. In this case, you have a *negative ion*.

Oxygen is an element that is ionized relatively easily. From here on, *ions* refer to oxygen ions, unless otherwise specified.

Positive ions and negative ions frequently occur randomly in the molecules of the air, because loose atmospheric atoms are constantly moving about, and sometimes bump into one another, dislodging an occasional electron. As you might suspect, there are usually an equal number of positive ions and negative ions. This equilibrium occurs because when atom A picks up an extra electron from atom B, atom A becomes a negative ion and atom B becomes a positive ion.

Under some atmospheric conditions, there might be a preponderance of either negative ions or positive ions in the air. For example, just before an electrical storm, there tend to be an excess of positive ions in the air. After a thunderstorm, the atmosphere has a surplus of negative ions.

Lightning has a very strong effect on air ionization. Lightning is just a powerful electrical discharge in the atmosphere, producing a lot of free electrons that can combine with nearby neutral oxygen atoms to produce negative ions.

There is a growing body of evidence that suggests human beings (and other animals) respond in various ways to ionization of the atmosphere. Perhaps you might have noticed feeling a little tense and edgy just before a big thunderstorm, but after the storm passes, you feel somewhat uplifted. (Some people are more sensitive to such effects than others.)

Excess positive ions in the atmosphere appear to cause tension and anxiety, and moderately heavy concentrations of negative ions in the atmosphere tend to enhance relaxation and feelings of well-being.

Of course, whenever people are tense and anxious, problems and conflicts are more likely to arise. The risk of accidents grows, and there is an increase in violent crimes and fights. Some studies suggest an increase in crime of up to 20% due to such atmospheric effects.

Persons subjected to moderately heavy concentrations of positive ions in the laboratory have difficulty relaxing, and some develop physiological symptoms, such as nausea or migraine headaches.

You should be aware that there is still some controversy in the scientific world about the effects of air ionization, but these theories seem to be gaining wide acceptance as more evidence is collected.

The negative effects of positive ions are more strongly established than the alleged benefits of negative ions. However, it has been experimentally demonstrated that exposure to negative oxygen ions can cause *serotonin* (a stress-related hormone) to break up into apparently harmless by-products.

Many people believe very strongly that running a negative ion generator (usually called an air ionizer) will have many significant health benefits. Many of these benefits are still unproven (though many of them appear to be likely to be proven soon).

One health benefit of negative air ionization has been pretty well established: clean air is obviously healthier than dirty air, which can be filled with smoke, pollen, and dust particles. Negative ions can actually help clean such foreign particles out of the air, by acting like tiny magnets for dust and other airborne particles. A negative ion will tend to attach itself to such a particle, weighting it down so that the particle sinks down to the ground, instead of hanging around in the air, waiting for someone to breathe it in. Because of this effect, negative ion generators are also frequently called air cleaners.

Special high-voltage circuits have been designed to emit negative ions into the air. Such a device is called an air ionizer, a negative ion generator, or an electronic air cleaner. Of course, commercial units are often given specialized names at the whim of the manufacturer.

All such devices work in a similar way. A high voltage is built up in some kind of probe. When this voltage is quickly discharged, it emits a strong burst of electrons, which attach themselves to oxygen atoms in the air surrounding the probe, thus creating many negative ions. In effect, a negative ion generator acts like a small, self-contained, and controllable source of pseudo lightning bolts (which are much safer than the real thing, of course).

The probe of an air ionizer is usually a thin needle or similar narrow, metallic object. The probe is given such a strong negative electrical charge (typically thousands of volts) that the excess electrons building up in the probe must eventually escape into the surrounding atmosphere, forming negative ions in the immediate vicinity. These ionized atoms soon spread out into the atmosphere.

If you place your hand near the probe while the air ionizer is operating, you might be able to feel the *ion wind* near the tip. This ion wind is due to the rapidly moving ion flux from the generator. Be very, very careful if you try to feel the ion wind for yourself. DO NOT EVER TOUCH THE PROBE WHILE THE NEGATIVE ION GENERATOR IS IN OPERATION. This is vitally important. Remember, the probe is at a very high electrical potential. A dangerous, possibly even fatal electric shock can almost be guaranteed if you touch the live probe!

Some air ionizers have the probe out in the open and fully exposed. This permits the device to exert its strongest effect, naturally enough, by providing more direct exposure to the open atmosphere. But, in some respects, this is an inherently risky and foolish design. In the interest of safety, a well-designed negative ion generator (especially for home use) should be housed in a way that it would be impossible, or at least very difficult for anyone to touch the charged probe during operation. For instance, the probe might be placed under a perforated plastic dome, as shown in Fig. 4-8. Plenty of air atoms can easily get in and out through the small holes in the dome, but curious fingers can't.

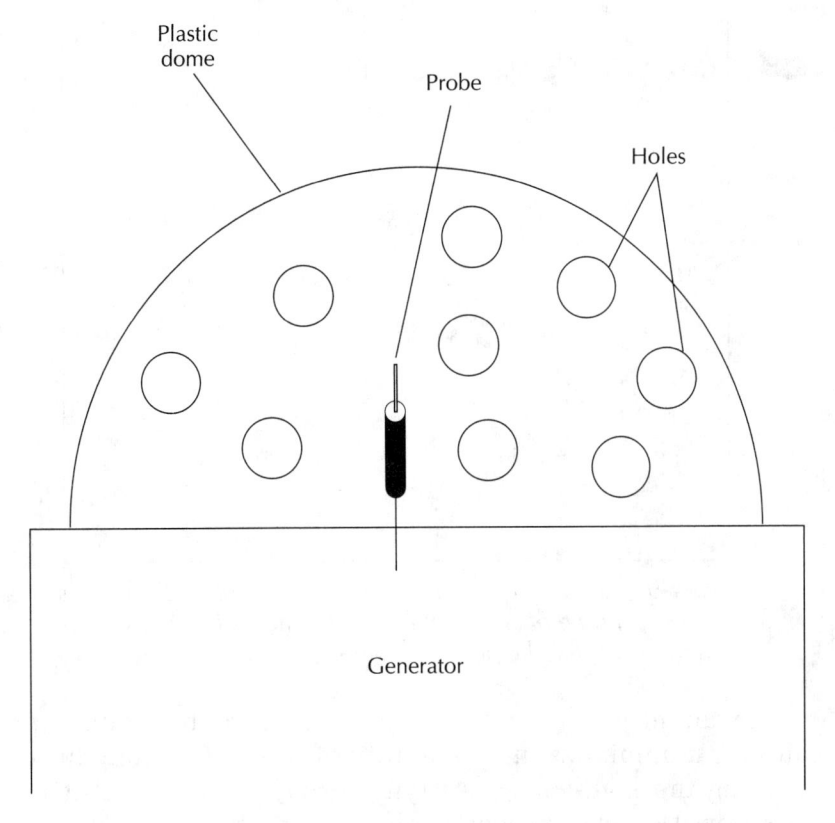

Fig. 4-8 *One way to shield the potential dangerous probe of an air ionizer is to place it under a perforated plastic dome.*

Another common approach is to recess a cavity in the main housing to partially enclose the probe, as shown in Fig. 4-9. The dimensions of the recessed cavity should be large enough to permit adequate air flow, but too small for fingers (especially a child's fingers) to fit inside.

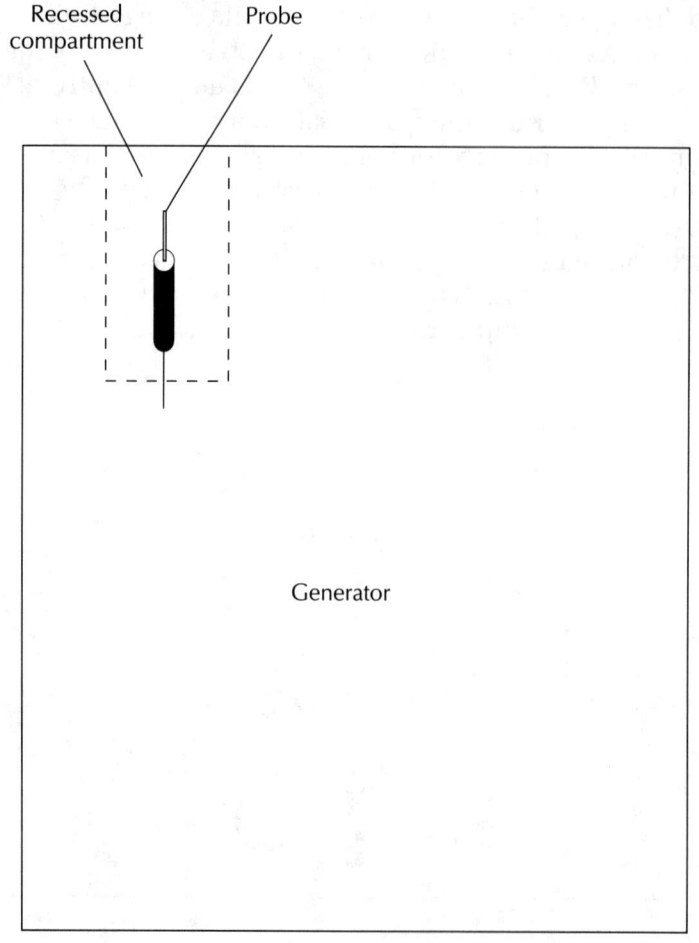

Fig. 4-9 *Another way to shield the potential dangerous probe of an air ionizer is to place it in a narrow recess in the project cabinet.*

In your air ionizer project, you should take the required precautions to minimize the possibility of someone getting hurt by touching the high-voltage carrying probe while the unit is in operation. Do not, however, enclose the probe entirely, or there will not be sufficient air flow for the negative ion generator to do any good. The entire project would be an utter waste of time and money in this case.

An air ionizer project should not present a significant shock hazard if you just use a little common sense and make reasonable provisions for safety.

In operation, the negative ion generator should be placed several feet away from the people who want to enjoy the benefits of the ion-enriched atmosphere provided by the generator. Obviously, this has the fringe benefit of limiting the possibility of any shock hazard—you can't touch what is out of your reach. More important, a little distance from the air ionizer will tend to maximize the positive benefits, because of the way air will typically circulate in a room.

Ideally, especially if children might be in the area, you should mount the negative ion generator high enough so no one can accidentally or easily get their hands on it while it is in operation.

Not surprising, an air ionizer will tend to work better in a partially enclosed room. But good ventilation in the room is strongly recommended. The room should be partially enclosed, but not fully insulated from the outside atmosphere.

This is not a suitable project for an electronics novice. Only experienced electronics experimenters should attempt building a project of this type. Remember, there are some very high voltages flowing throughout the circuitry in this project.

Figure 4-10 shows the circuit for your air ionizer project. A suitable parts list for this project appears in Table 4-5.

Because of the high power levels that must be handled by the circuit, most common electronic components will not be suitable for use in this sort of project. Use of special components is unavoidable, considering the nature of the circuit you are trying to build. You might experience some difficulty in locating some of the parts you need for this project. Often your best bet will be to check with local electronics jobbers who deal with industrial equipment. They might be willing to sell you a few special parts for a reasonable price. Another good source would be to check the electronics surplus houses that advertise in the backs of the electronic hobbyist magazines. Most large cities will also have a few local electronics surplus dealers as well.

Save yourself some potential frustration, and make sure you have an available source for all of the required components before investing any money into the project. It is a good idea to be sure you can get all the needed components before beginning any electronics project, but it is particularly advisable when the circuit calls for several unusual components, as is the case here.

158 Atmosphere-related projects

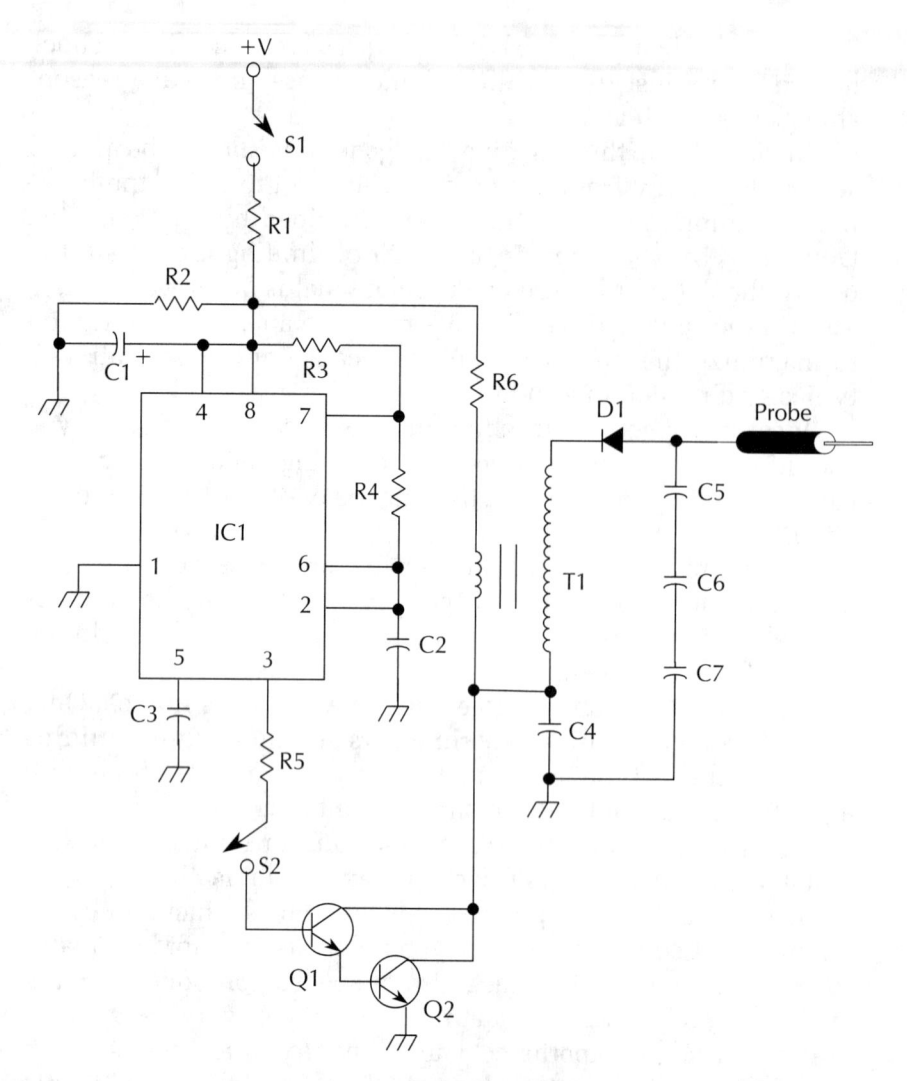

Fig. 4-10 *Project 24. Air ionizer.*

The circuit itself is fairly simple. It is basically just a high-voltage pulse generator. The system used here is known as the *high-voltage corona discharge method,* and it is probably the most commonly used approach in designing negative ion generators today.

IC1 is a 555 (or 7555) timer chip, operated in the astable mode. The 7555 is a CMOS (complementary metal-oxide semiconductor) version of the 555, and it is pin-for-pin compatible. No changes in the circuitry are required to substitute a 7555 for

Table 4-5 Suggested parts list for Project 24. Air ionizer.

IC1	7555 or 555 timer
Q1	NPN transistor (2N2102, Radio Shack RS2030, or similar)
Q2	NPN transistor (GE-19, SK3027, MJE3055, Radio Shack RS2041, or similar)
D1	45 kV diode array (ECG513 or similar)
T1	automotive ignition coil (12 V, 3 terminal)—see text
C1	500 µF 35 V electrolytic capacitor
C2, C4	0.047 µF capacitor
C3	0.01 µF capacitor
C5, C7	470 pF 6000 V capacitor
C6	330 pF 6000 V capacitor
R1	1 kΩ 10 W 10% resistor
R2	10 Ω 10 W 10% resistor
R3	390 kΩ ¼ W 5% resistor
R4	39 kΩ ¼ W 5% resistor
R5	1 kΩ ¼ W 5% resistor
R6	1.8 Ω 10 W 5% resistor
S1, S2	SPST switch

a 555. The 555 timer is one of the most popular and widely available of all ICs on the market today. Using the component values suggested in the parts list, the pulse frequency generated by IC1 will be about 65 Hz, with a duty cycle of a little less than 10%. The small duty cycle figure means the circuit generates a string of very narrow pulses.

Transistors Q1 and Q2 make a Darlington-pair amplifier for boosting these pulses. Both must be NPN transistors. The exact type numbers aren't critical (they don't have to be identical), but Q2 must be a heavy-duty power transistor. These transistors must conduct a lot of heavy power (especially Q2), so by all means use adequate heat sinking. When in doubt, use more heat sinking. The only possible disadvantages of too much heat sinking are purely aesthetic. It's worth a little extra cost and circuit bulk to be on the safe side. You certainly don't want your project to self-destruct as it is operating.

The output from the amplifier stage is T1, which is actually a standard 12 V, three-terminal automobile ignition coil. You probably won't be able to find anything directly suitable at an electronics parts store, but you can readily find such ignition coils at an automobile parts supply store, or you might find a nice bargain at an automotive junk yard. Exact specifications for this device are not critical, but it must be of the three-terminal type.

If you are not familiar with automotive ignition coils, you might have a little difficulty locating the high-V+ output terminal. That is because in its normal use (in an automobile engine) the connection is made with a heavy-gauge spark plug wire clip. The high-voltage terminal is inside a small tubelike protrusion at the top of the ignition coil unit. The stripped end of a heavy-gauge, well-insulated wire should be firmly inserted into the opening of the protrusion. Make sure there is a good, strong mechanical connection that won't pull free too easily. Except for the ends making the actual electrical connections, this wire should be very thickly insulated. Remember, it will be carrying a dangerously high level of electrical power. There is no such thing as too much insulation in this application.

It would be a very good idea to completely enclose the entire ignition coil assembly in a heat-resistant plastic (or otherwise insulated) case of some sort for maximum protection against shock hazards. The ignition coil (T1) is much more of a risk for electrical shock than the ionization probe itself.

The coil assembly boosts the potential of the pulses considerably. The signal is then half-wave rectified by D1, which is not an ordinary diode, but a 45 kV (45,000 V) high-voltage diode assembly. This will be one of the harder to find components in this project, as hobbyist projects rarely work with such high-voltage circuits. But such devices are in widespread industrial use. They are also used in some television sets and other consumer products.

Make the connections to the diode assembly via standard high-voltage ignition wire. You will definitely need a fairly hefty wire with good (and thick) insulation.

Of course, it is always important to be careful about the polarity when installing components such as diodes, but it is particularly critical for D1 in this project. If this diode assembly is installed backwards, your project will become a positive ion generator, which almost certainly would not be at all desirable, as suggested by the general discussion that began this section.

Filter capacitors C5 through C7 must also be high-voltage units. The total series capacitance works out to a little under 140 µF. Remember, the formula for capacitors in series is:

$$\frac{1}{C_T} = \frac{1}{C_1} + \frac{1}{C_2} + \frac{1}{C_3}$$

The reason a single 150 µF capacitor is not used is because of the very, very high voltage at this point in the circuit. It would be extremely hard to find a single capacitor that could bear the full output voltage of the generator. By using three high-voltage capacitors, they can effectively share the power load. In this case, three capacitors are almost certainly less expensive than one to do the same job.

These three capacitors working together act pretty much like a standard filter capacitor in any half-wave power rectification circuit. Once filtered, the high-voltage pulses from the generator circuit are fed to the probe. A common sewing needle is a good choice for the probe. It is an almost ideal size and shape. The probe should definitely have a pointed tip at the top.

Although this circuit generates a very high voltage (about –6 kV to –9 kV or –6000 V –9000 V), the actual output current is relatively low, so if someone does accidentally touch the probe, it probably won't be fatal. Of course, there is no way to guarantee this. Whenever high power levels are involved, there is always some risk of potentially fatal electrical shock, especially for anyone with any kind of heart ailment. At the very least, the probe most definitely provides a very substantial kick, and can give you a very painful shock. It could cause you to fall or jerk back and hurt yourself on some nearby object. Be careful with this project. It is designed to be as safe as it can be, but you must treat it with all due respect for the high voltages involved.

Of course, the entire circuit must be completely enclosed in a fully insulated case. Once again, you can't use too much insulation. When in doubt, use more.

It is strongly recommended that the probe be shielded or recessed, as discussed earlier in this section. But remember, for the project to function usefully, there must be adequate air flow around the tip of the probe.

Never use this project around children if they might be unsupervised, even for a moment. Children can be very curious and very clever. They can often find ways to defeat the most elaborate and carefully thought-out "foolproof" safety devices.

As an added precaution, two power switches are included in this project. Close switch S1 first to apply power to the pulse generator (IC1), then close switch S2 to permit the pulses to go through the amplifier and the ignition coil (T1) to the probe for the release of negative ions. If you have any children in the house, use a locking key switch for S2, and keep the key secluded in a safe place.

If you decide to build this project, please do yourself a big favor and reread all of the safety precautions in this section carefully.

There is no reason in the world for anyone to be hurt by a project like this. Please, please don't ever take any foolish chances.

❖ 5
Light-related projects

Another important part of the environment is light. Light is a form of electromagnetic energy. Only a fairly narrow band of frequencies in the total light spectrum are visible to the human eye. The exact frequency determines the color seen. The lowest visible frequency looks deep red. At the highest visible frequency, the perceived color is violet.

Even though your eyes can't see them, there isn't much difference between the visible light frequencies, and invisible light at nearby frequencies. For example, energy in the frequency band lower than red (the lowest visible frequency) is called *infrared light*. Similarly, *ultraviolet light* occurs at frequencies just above the visible spectrum.

Most standard photosensors respond primarily to frequencies in the visible spectrum, pretty much like the human eye. Some specialized photosensors that respond to infrared light frequencies are also available. If it suits your application, you can easily substitute an infrared sensor for the visible light sensors assumed here. You shouldn't need to make any other changes in the circuitry of any of these projects to support an infrared sensor of the appropriate type.

Some of the light-related projects presented in this chapter detect the light intensity in the immediate environment and respond accordingly. Other circuits are designed to control lighting devices in the environment. A few of the projects in this chapter do both of these things.

Light-operated relay

Virtually any electrically powered device can be controlled with this project, provided you use a hefty enough relay. The relay is activated by the presence of light of a sufficient intensity.

The schematic diagram for this light operated relay project is shown in Fig. 5-1. A typical parts list for this circuit is given in Table 5-1. Nothing is terribly critical here. You shouldn't have any difficulty at all locating the required components for this project.

Fig. 5-1 Project 25. Light- operated relay.

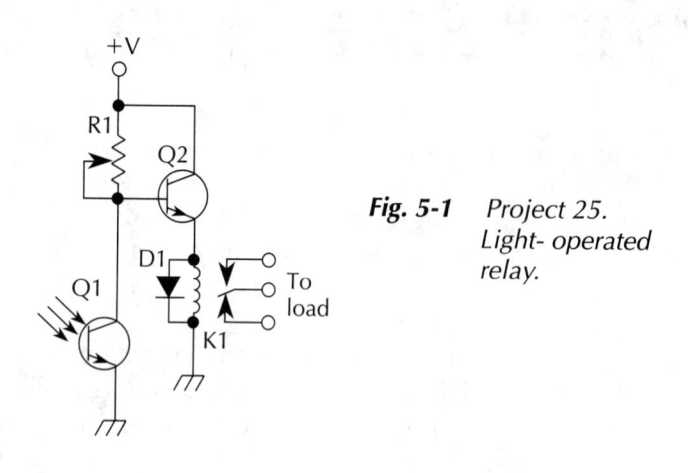

**Table 5-1
Suggested parts list for
Project 25. Light-operated relay.**

Q1	NPN phototransistor
Q2	NPN transistor (2N3904 or similar)
D1	diode (1N4002 or similar)
K1	9 V relay (500 Ω coil) (contacts to suit load)
R1	100 kΩ potentiometer

A phototransistor (Q1) is used as the sensor for this circuit. When the light striking the lens cap of the phototransistor exceeds a specific level, the relay is activated. When the sensed light level drops below the trigger point, the relay is deactivated. The trigger point, or the sensitivity of the circuit, can be set with potentiometer R1.

Transistor Q2 is a simple amplification stage to boost the output signal from the phototransistor (Q1) sufficiently to activate the relay. Almost any low-power NPN transistor should work just fine in this circuit. The requirements placed on this transistor are very minimal. It doesn't have to amplify the signal very much, and it doesn't even matter if the signal is badly distorted or moderately noisy. As long as the amplifier output signal goes high enough to reliably activate the relay, that's all that matters in this project.

Diode D1, as usual, is included to protect the delicate relay coil from possible damage cause by back EMF during switching. Almost any standard diode should do the job.

This circuit should work reliably with supply voltages anywhere from +6 to +15 V.

For many applications, some sort of light shield might need to be mounted over the sensor (Q1) to reduce false triggering problems from uncontrolled ambient light sources.

The relay should be selected so that it's switching contacts can safely handle the desired load. If you need to control a very high current load, you can use one relay to control a second, larger relay, as shown in Fig. 5-2.

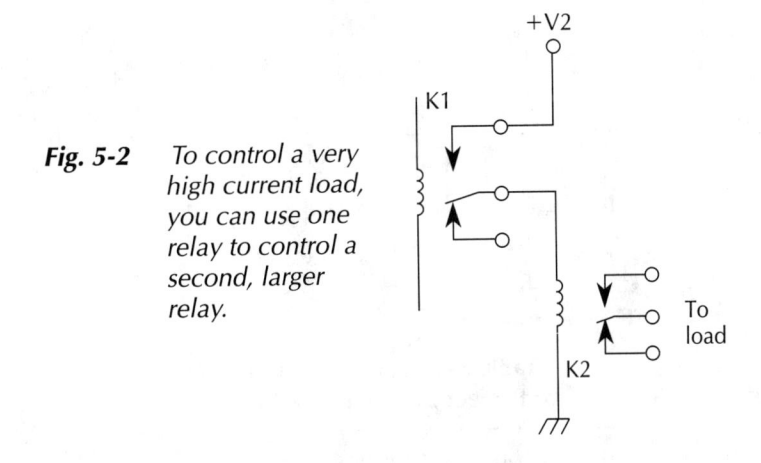

Fig. 5-2 To control a very high current load, you can use one relay to control a second, larger relay.

Dark-operated relay

The next project is sort of a mirror image of the previous project. It operates in exactly the opposite way, which can be more useful in certain applications. This time the relay is activated when the relay is dark, rather than when it is illuminated.

The circuit for the dark-operated relay project is shown in Fig. 5-3. A suitable parts list for this project is given in Table 5-2.

Again, nothing is terribly critical here. You shouldn't have

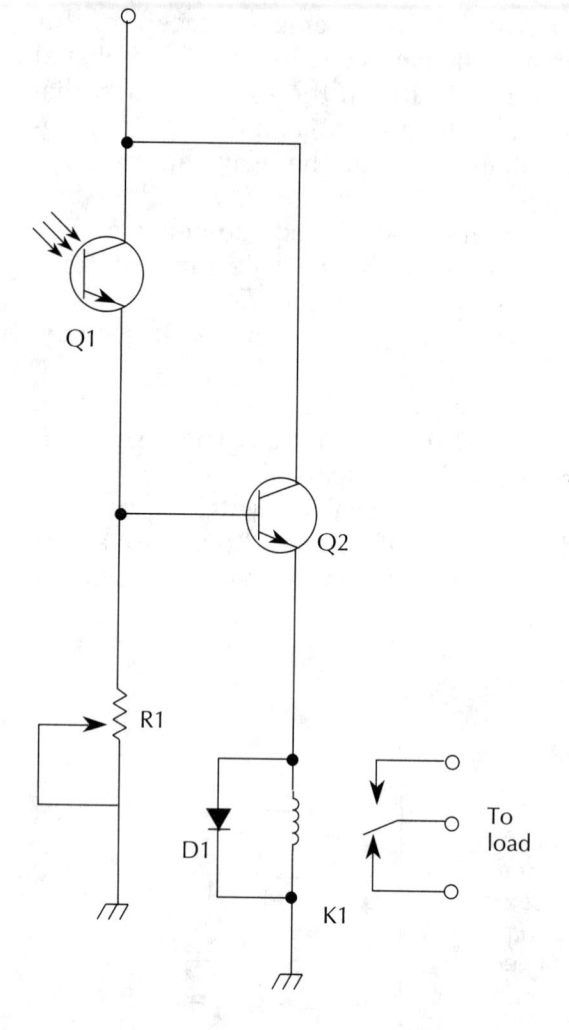

Fig. 5-3 Project 26. Dark-operated relay.

Table 5-2 Suggested parts list for Project 26. Dark-operated relay.

Q1	NPN phototransistor
Q2	NPN transistor (2N3904 or similar)
D1	diode (1N4002 or similar)
K1	9 V relay (500 Ω coil) (contacts to suit load)
R1	100 kΩ potentiometer

any difficulty at all locating the required components for this project. In fact, this circuit is very similar to the preceding project, except for the relative placement of the phototransistor sensor (Q1) and the sensitivity control (potentiometer R1).

As in the earlier project, transistor Q2 is a simple amplification stage to boost the output signal from the phototransistor (Q1) sufficiently to activate the relay.

Diode D1 as usual is included to protect the delicate relay coil from possible damage caused by back EMF during switching. Almost any standard diode should do the job.

For many applications, some sort of light shield might need to be mounted over the sensor (Q1) to avoid reliability problems from uncontrolled ambient light sources. A light shield will especially be necessary in some well-lit environments, or the relay might not be activated. Why would you want to use a dark-operated relay circuit in a well-lit area? It can detect shadows passing over it, such as when a person (or any opaque object) passes between the sensor and the light source.

As long as the light shining on the sensor (phototransistor Q1) is above the trigger level (as set by R1), the relay will be deactivated. When the light intensity drops below the trigger level, or when the sensor is shaded from the light source (by any means), the relay is activated. The relay should be selected so that it's switching contacts can safely handle the desired load.

This circuit should work reliably with supply voltages anywhere from +6 to +15 V.

Self-activating night light

In certain areas, a night light can be very helpful. But it is wasteful and inelegant to leave it burning all the time. But it is often too easy to forget to turn it on and off, and it might be awkward to try to find the switch in the dark. This project is a simple circuit for a night light that automatically turns itself on when it gets dark, then back off again when the ambient light increases sufficiently. The light is only turned on when it is needed.

The schematic diagram for the self-activating night light circuit is shown in Fig. 5-4. A suitable parts list for this project is given in Table 5-3. Nothing is very critical here, and only common, easy-to-find components are called for.

Virtually any NPN phototransistor should work for Q1 in this circuit. For some phototransistors, a different value for resistor R1 might help provide a more convenient range for the sensitivity control. Experiment with the circuit on a solderless breadboard.

The SCR (Q2) should be selected to suit the load. In most applications, this will be a relatively small night light. A low-

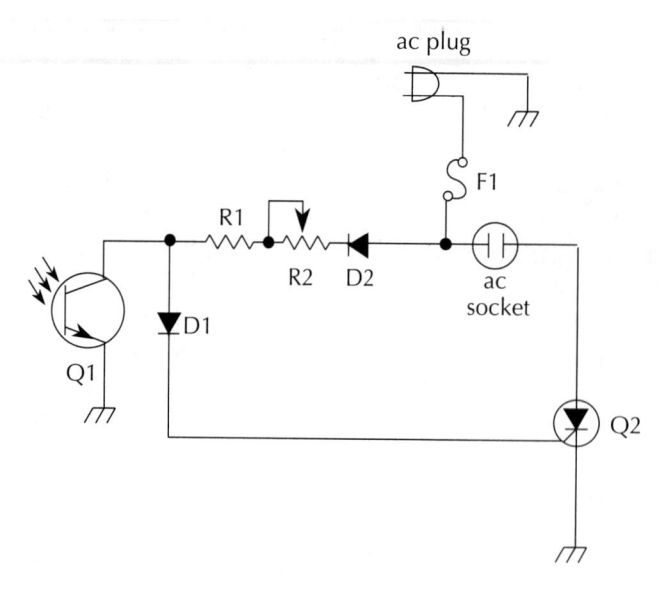

Fig. 5-4 *Project 27. Self-activating night light.*

Table 5-3 Suggested parts list for Project 27. Self-activating night light.

Q1	NPN phototransistor
Q2	SCR (select to suit load)
D1, D2	diode (1N5059 or similar)
R1	1 MΩ ¼ W 5% resistor
R2	5 MΩ potentiometer
F1	fuse to suit load and SCR
ac socket	
ac plug	

power bulb will probably be used, so a very low-power SCR will usually be suitable. As always, it is a good idea to overrate the SCR slightly, so it can safely handle more current than the maximum current level that you expect the load to draw. This gives you a little headroom for protection against the unexpected. If the load is large, using a heat sink is always a good idea with an SCR.

This project is ac powered, which means the circuit must be completely insulated and shielded. It should be impossible for anyone to touch any circuit connection point (including those nominally at ground potential) at any time while the project is receiving power. Take all normal safety precautions.

The fuse (F1) should be selected to suit the load. The control circuit itself consumes only negligible current. Do not omit or overrate the fuse. That could be very dangerous. Saving a few pennies on a fuse is never worth the risk. Double check all wiring very carefully before applying power to the circuit.

This project is quite straightforward in its operation. The phototransistor (Q1) serves as the light sensor. As long as it is sufficiently illuminated, the phototransistor shunts the SCR (Q2) gate current to ground, and nothing happens. The load receives no power.

When the ambient lighting level drops below a specific trigger point, however, the phototransistor sends current into the gate of the SCR, and the SCR is switched on. While the SCR is conducting, it permits current to flow through the ac socket and whatever load (the night light) is plugged into it.

Potentiometer R2 sets the sensitivity of the detector. Adjust this control for the desired turn-on point. Expose the phototransistor to the light level you want to turn on the night light. Adjust R2 so the load is fully off, then turn it slowly until it just turns on. This is all there is to calibrating this project.

In some applications, you might want to add a manual reset switch across the anode and cathode of the SCR (Q2). This should be a normally open SPST momentary contact push-button switch.

Light dimmer

Lamp dimmers are always popular projects. They permit you to adjust the amount of light in the room to suit your needs or your moods. For example, brighter light is suitable for reading, and a dimmer light might be desirable for a romantic dinner.

A very simple, but effective light dimmer circuit is shown in Fig. 5-5. The parts list for this project is given in Table 5-4. Nothing is particularly critical in this project. You have considerable practical leeway in selecting the components you use in this project. The most critical component in this circuit is the triac (Q1), which is the heart of this project. This triac must be selected for an adequate current handling capability for the load in your intended application. Add a little elbow room for the sake of safety and to protect against unanticipated power surges. For example, if you intend to control a 2 A load with this project, use a triac rated for at least 3 A.

No matter what triac you select, don't use this project to control a load (lighting device or devices) that is more than about 400 W (watts).

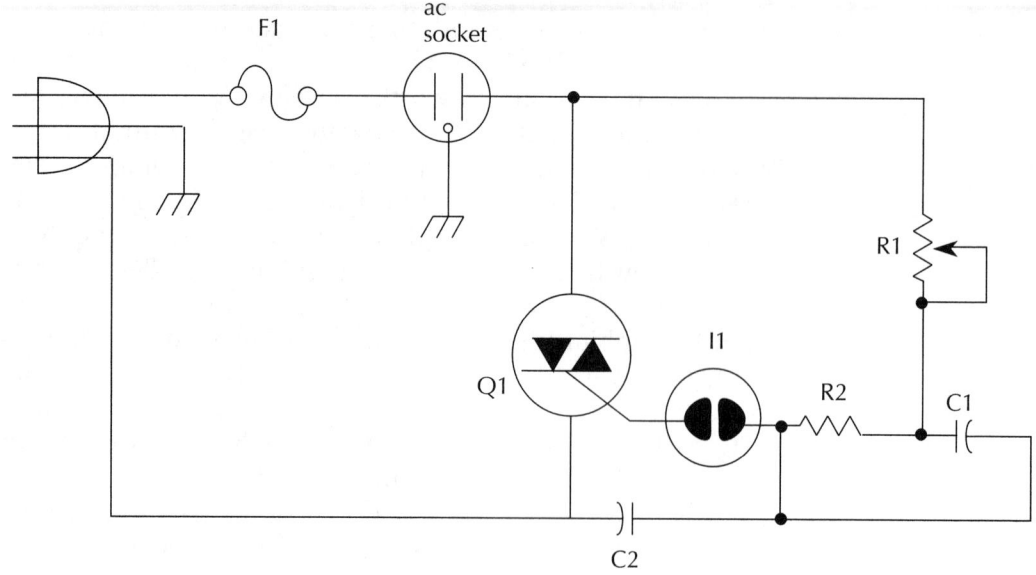

Fig. 5-5 Project 28. Light dimmer.

Table 5-4 Suggested parts list for Project 28. Light dimmer.

Q1	Triac to suit load (RCA 40502 or similar)
I1	NE-2 neon lamp
F1	fuse to suit load
C1, C2	0.68 mF 250 V capacitor
R1	50 kΩ potentiometer
R2	15 kΩ ½ W 5% resistor

For most relatively low-power applications, a heat sink is not absolutely required for the triac in this project, especially if the device is sufficiently overrated for the intended load. But adding a heat sink certainly wouldn't hurt and could help prevent premature circuit failure.

Remember, this project is intended only for the control of electrical lights and similar devices. Do not attempt to use this circuit to control an inductive load, such as an electrical device involving a motor. This circuit cannot be used as a speed control for an electric drill, or any similar application.

In operation, potentiometer R1 is used to control the amount of power reaching the output socket and therefore the brightness of the lamp plugged into the socket.

There is a minor but important compromise involved in this circuit. The neon bulb will not trip the triac until it conducts enough to turn the output lamp on at a moderately bright level. This means there is a dead space in the operation of the circuit at very low power levels. Potentiometer R1 needs to be turned past this critical point to turn the lamp on, then it can be backed off to a somewhat dimmer glow, if desired.

Light cross-fader

The next project is designed to control two electrical lights in opposition automatically. As one is increased in brightness, the other is dimmed. In other words, it permits you to easily perform a cross fade from one light to another. This project is not recommended for use with any electrical load other than lighting devices. Especially avoid inductive loads, such as electric motors, which could be severely damaged by the control signals generated by this circuit.

The schematic diagram of the cross-fader circuit is shown in Fig. 5-6, and the parts list is given in Table 5-5.

Nothing is very critical in this circuit. The most important components are the two SCRs (Q1 and Q2), which should be selected to suit the intended load for your application. Overrate the SCR current-handling capability. For example, if the load will draw up to 1 A, it would be a good idea to select an SCR that can handle at least 1.5 A.

It is always advisable to use a fuse of suitable value in series with any ac power output. Add fuses (not shown) to protect ac socket 1 and ac socket 2 in Fig. 5-6. Because this circuit uses ac power, it should also have an input fuse (F1).

Initially one of the controlled lamps will be fully on and the other will be fully off. When this circuit is triggered, the first lamp will gradually be faded down to full off, as the other one is simultaneously faded up to full on. If both lamps have the same wattage, and they are closely placed, the overall lighting level should remain fairly constant during the cross-fading process.

In some applications, you might want to use filtered lamps, which glow with differing colors. The effect of a two-color cross-fade can be quite impressive.

You might want to experiment with alternate values for resistor R1 and capacitor C1 to create different time periods for the cross-fade.

172 Light-related projects

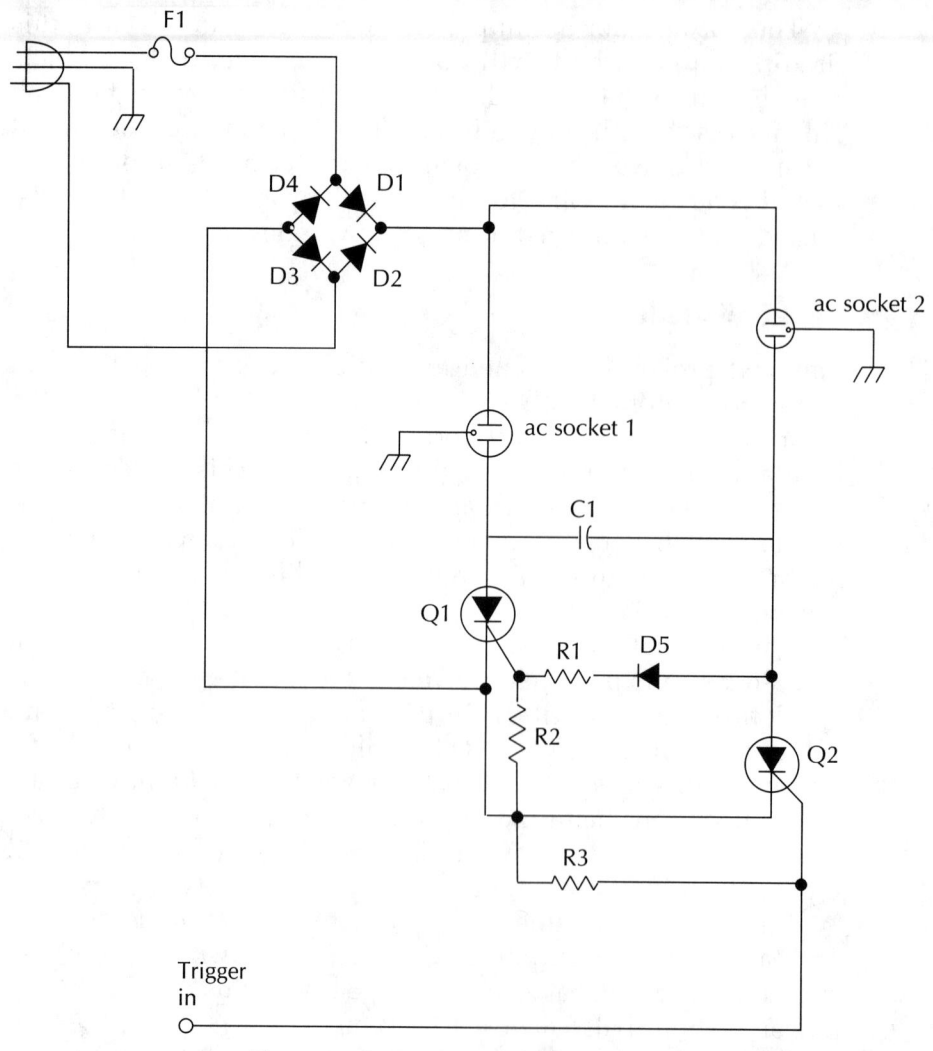

Fig. 5-6 *Project 29. Light cross-fader.*

Table 5-5 Suggested parts list for Project 29. Light cross-fader.

Q1, Q2	SCR to suit load (C106B or similar)
D1–D5	diode (1N4003 or similar)
F1	2 A fast-blow fuse
C1	0.47 µF 250 V capacitor
R1	47 kΩ ½ W 5% resistor
R2, R3	4.7 kΩ ½ W 5% resistor

Sequential light controller

A rather novel type of automated lighting control circuit is shown in Fig. 5-7. A suitable parts list for this project is given in Table 5-6.

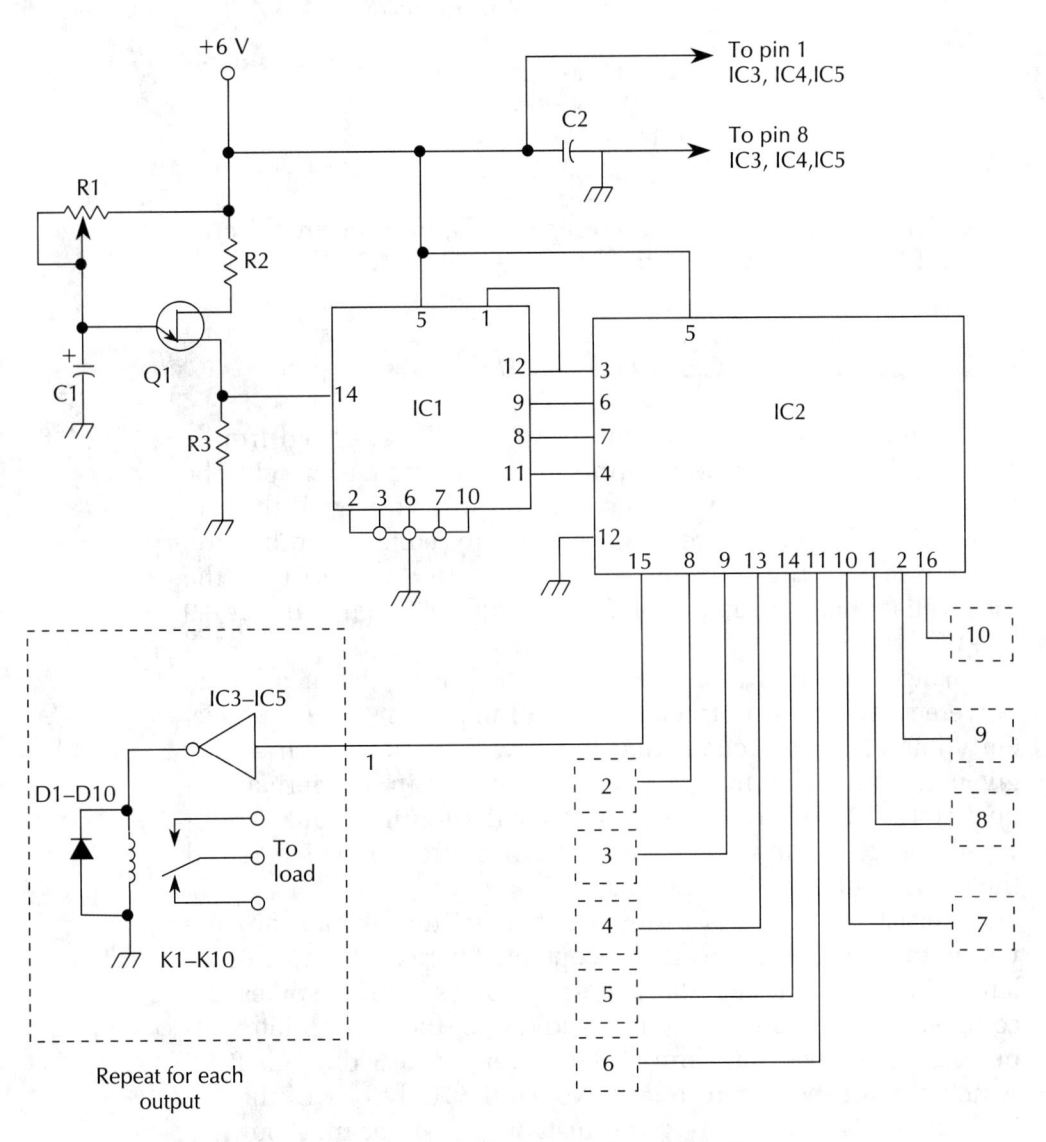

Fig. 5-7 *Project 30. Sequential light controller.*

Table 5-6 Suggested parts list for Project 30. Sequential light controller.

IC1	74C90 decade counter
IC2	74C91 BCD-to-decimal decoder
IC3, IC4, IC5	CD4049 hex inverter
D1–D10	diode (1N4002 or similar)
K1–K10	relay to suit loads
C1	4.7 µF 25 V electrolytic capacitor
C2	0.1 µF capacitor
R1	1 MΩ potentiometer (sequence rate)
R2	1 kΩ ¼ W 5% resistor
R3	100 Ω ¼ W 5% resistor

CMOS ICs are recommended for all ICs, but you might substitute TTL chips, if you prefer:

IC1	7490
IC2	7491
IC3, IC4, IC5	7404

Of course, the control circuit must then be operated from a well-regulated +5 V power supply. TTL devices can quickly be damaged or destroyed by an incorrect supply voltage. If the recommended CMOS ICs are used, the supply voltage can be anything from +6 V up to about +15 V. Generally, you will get the most reliable circuit operation if the supply voltage is in the +9 to +12 V range.

Using this circuit, up to ten independent lights (or other ac powered devices) are turned on and off in a repeating sequence. Only one of the ten controlled lights will be turned on at any given instant. The lights will be operated in numerical sequence (1 - 2 - 3 - and so forth). After the tenth output is activated, the next step in the sequence will return to device 1, and the entire cycle will be repeated.

Transistor Q1 is a UJT (unijunction transistor). Almost any device of this type should work. It is operated here as a simple oscillator circuit, producing sharp output spikes. These spikes are counted and processed by IC1 and IC2, then the appropriate output signals are fed out through ten inverters (IC3 through IC5), which control the output relays. Notice that only one of the ten IC2 outputs is LOW at any given instant. The other nine outputs will be held HIGH. The inverters reverse the output states to activate the appropriate relay for each step in the control sequence.

Adjusting potentiometer R1 will determine the switching speed of the sequence. This might be a trimpot or a front panel potentiometer, depending on the specific requirements of your individual intended application. To change the overall range of available sequence rates, you might try experimenting with alternate values for capacitor C1.

It would be a good idea to include a 0.01 µF capacitor across the power supply pins (pin 1) of IC3, IC4, and IC5. This will protect the chips from possible noise surges riding on the supply voltage for whatever reason.

It would also be a good idea to include an appropriate fuse in each of the ten output circuits. Of course, the relays (K1 through K10) should be selected so their switch contacts can safely withstand the intended current loads. Drive only relatively low power (100 W or less) electrical devices from the outputs of this sequential control circuit.

This is a fun project. Possible applications include eye-catching displays or warning devices. In some applications, you might want to physically arrange the lamps out of numerical order. This can almost give a pseudorandom effect under some circumstances. In other applications they should be positioned in strict numerical order, which will make the sequential pattern the most obvious.

At moderately high sequence frequencies, if the controlled lamps are relatively closely placed, you will get a light-chaser effect. It won't look so much like one light going off and the next coming on, as a continuously moving light source.

Automated guest greeter

Perhaps you don't want to leave a porch light on every night just in case you might have an unexpected guest after dark. It would not be a major waste of energy, but some energy would be consumed to no good purpose. On the other hand, you don't want any such guest to have to wait outside in the dark until you can answer the door. The circuit shown in Fig. 5-8 offers a rather clever solution to such situations. Whenever the doorbell is rung, the porch light will be automatically turned on for a predetermined period of time, then it will shut itself off.

A suitable parts list for this project is given in Table 5-7. None of the component values are particularly critical, nor should you have any difficulty finding any of the parts needed to construct this project. You must have an electrical doorbell

176 Light-related projects

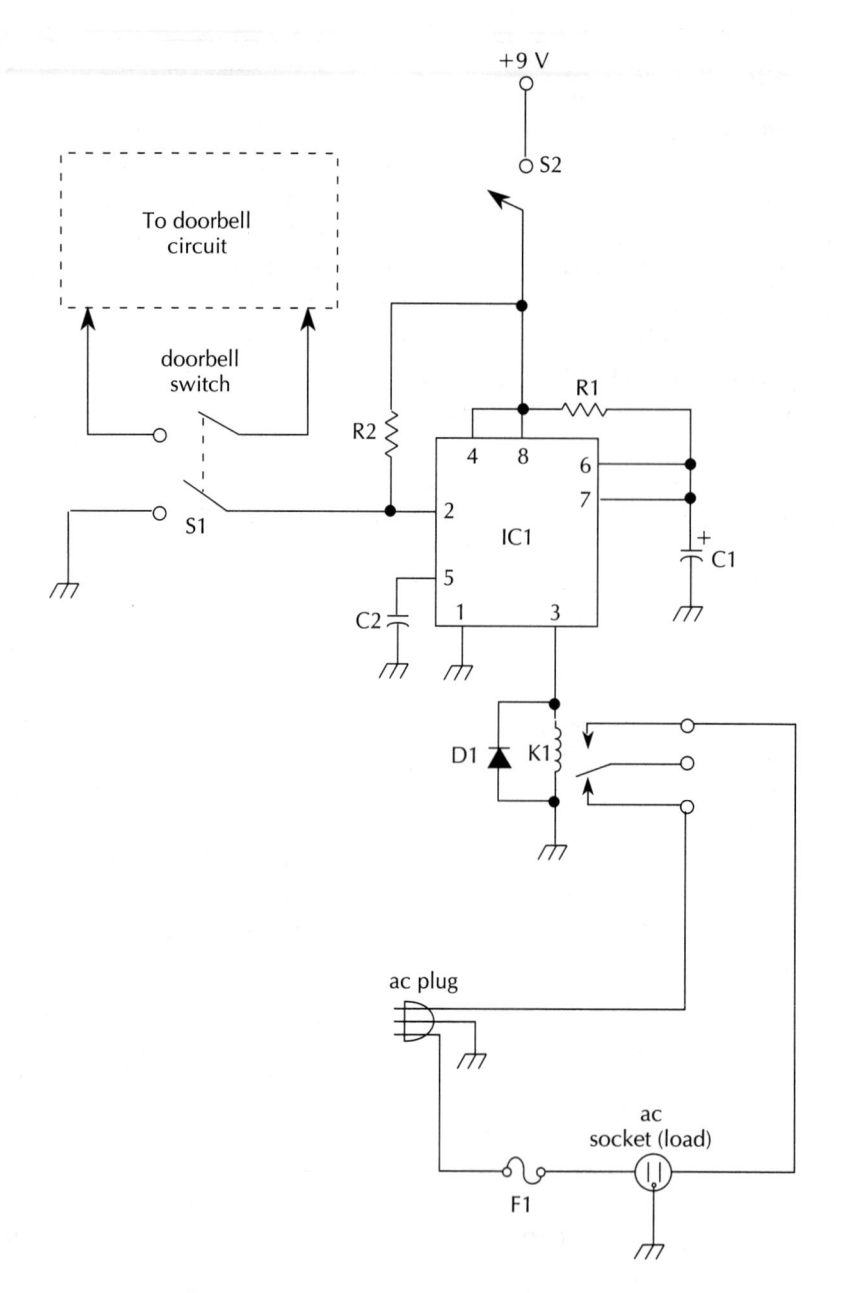

Fig. 5-8 *Project 31. Automated guest greeter.*

already installed of course. The project won't work if someone knocks on the door, or uses a mechanical doorbell.

Ordinarily, the doorbell is activated with an SPST switch. In this project, you simply replace the original doorbell switch with a DPST switch. DPST switches tend to be difficult to find, but you can use a DPDT switch just as well. Merely leave the

Table 5-7 Suggested parts list for Project 31. Automated guest greeter.

IC1	555 timer
D1	diode (1N4002 or similar)
K1	relay to suit load
F1	fuse to suit load
S1	normally open DPST push-button switch (doorbell switch—see text)
C1	250 µF 35 V electrolytic capacitor—see text
C2	0.01 µF capacitor
R1	1 MΩ ¼ W 5% resistor
R2	10 kΩ ¼ W 5% resistor

extra switch contacts disconnected, and as far as the electrical circuit is concerned, they won't exist at all. One section (or *pole*) of this new switch is connected directly to the doorbell in the usual manner. The operation of the doorbell itself won't be affected in any way. The second half (pole) of the new switch is used to send a trigger signal to a timer circuit.

The actual circuitry in this project is built around the popular 555 timer IC, wired in its monostable multivibrator mode. You can substitute the CMOS version of this device (7555), if your prefer. The 555 and the 7555 are pin-for-pin compatible, and such a substitution would require no changes in the external circuitry at all.

Most of the time the trigger input (pin #2) of the timer (IC1) is held HIGH through resistor R2. When the lower half of the doorbell switch (S1) is closed, pin 2 is grounded, which is electrically the same as a LOW signal. The 555 timer is triggered by a HIGH to LOW transition, so this will do the trick.

When triggered, the timer brings its output (pin 3) HIGH for the duration of a predetermined timing period. This HIGH signal activates the relay (K1), which permits power to be sent to the controlled device (light) connected to the ac socket.

This basic circuit can be used to control almost any electrically powered device. Use your imagination to come up with some additional applications. Anything that can operate a switch (manually or electronically) can trigger the timer circuit.

The timing period is determined by the values of resistor R1 and capacitor C1. The timing formula, as in any standard 555 monostable timer circuit, is:

$$T = 1.1 R_1 C_1$$

where T is the timing period in seconds, R_1 is the resistance in ohms, and C_1 is the capacitance in farads. Be sure to use the correct units, or you won't get correct results from the equation.

Using the component values suggested in the parts list, the timing period for the circuit will work out to about:

$$T = 1.1 \times 1,000,000 \times 0.00025$$
$$= 275 \text{ seconds}$$

Each time the doorbell is rung, the porch light will be turned on for approximately 4½ minutes (nominally 4 minutes and 35 seconds). Then it will shut itself off. This choice is a good one for the timing period in the automated guest-greeter period. By the time the circuit times out and turns out the light, you'll have had plenty of time to answer the door, or most guests will have given up and left.

Of course you should feel free to experiment with different component values for R1 and C1 to give different timing periods, especially if you are using this project for some other application of your own.

Select the value of fuse F1 to suit the intended load for your project. Don't use more than about a 2 or 2.5 A fuse in this circuit. This will permit a load of up to 240 to 300 W, which should be quite sufficient. For safety reasons, an appropriately valued fuse should always be used in any circuit carrying ac power.

The relay should also be selected so its switch contacts will carry more than the required load current. Diode D1 protects the delicate relay coil from possible back EMF spikes when it is switching states. Almost any standard silicon diode should work for this application.

Notice that ac power is carried only in the output portion of the circuit (from the relay switch contacts onward). The main body of the circuit (IC1 and its associated components) must be operated from its own dc supply voltage. Anything from +9 to +12 V can be used. You can either use battery power, or an ac-to-dc converter. In this particular application, the second option would be the most logical choice.

Switch S2 cuts off the dc power to the timer circuitry. This disables the circuit, but permits the doorbell to work normally. There would be no point in turning on the porch light for four and a half minutes whenever someone rings the doorbell in the daytime. IC1 will not be damaged by being triggered when it is not receiving any power. With switch S2 open, there will be no voltage through resistor R2, so the voltage at pin 2 will be 0.

When the doorbell switch (S1) is closed, pin 2 of IC1 will simply be grounded, which means it will still be getting an input voltage of 0. In other words, nothing at all happens.

This circuit is simple and elegant, and the basic concept of the project can easily be adapted to many other practical applications. Use your imagination.

Photosensitive automatic porch light

Naturally, there isn't much point in leaving a porch light on during the day. That just wastes energy—not much energy, of course, but why waste any you don't need to? But it is all too easy to forget to turn the porch light on at dusk, and off again when you go to bed, or in the morning. But you never know when evening guests might show up. It would be nice to have a light on for them.

The circuit shown in Fig. 5-9 is designed to automatically turn a light (or other electrically powered device) on at dusk, and off at dawn. The parts list for this project is given in Table 5-8.

Nothing is terribly critical in this project. There is plenty of room for reasonable substitutions. You shouldn't have any difficulty locating the components required for this circuit.

Notice that a 555 timer (IC1) is included in this circuit. This adds a little pseudohysteresis to the system. The purpose of the timer is to prevent the light from blinking on and off in response to a passing cloud or a moving shadow during the day, or a reflection, or headlights of a passing car at night. This would be annoying, and would shorten the lifespan of the light bulb(s) controlled by the circuit. The timer introduces a short delay before the control circuit responds to a significant change in the detected lighting level. Using the component values suggested in the parts list, the timer delay period will be a little less than two minutes. Assuming the components are exactly on value (0% tolerance), the time delay period would be 112 seconds long. If you choose, you can change the circuit time delay period by experimenting with different values for resistor R1 or capacitor C1.

The light sensor in this project is P1, a common photocell or solar battery. When illuminated, a single photocell will generate 0.5 V. The current will be determined by the physical size of the photocell's sensor surface, and the intensity of the light striking it.

This signal from the photosensor is boosted by a simple amplifier made up of transistors Q1 and Q2, wired as a Darlington

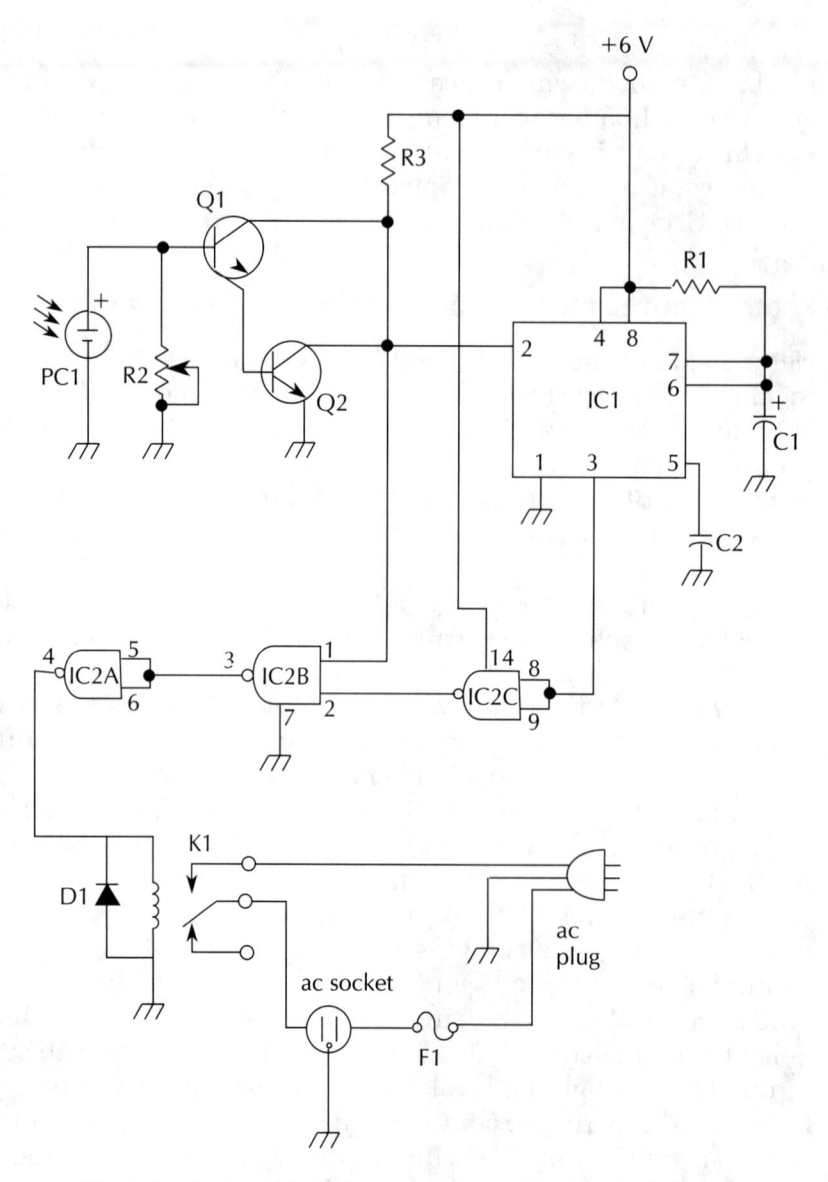

Fig. 5-9 *Project 32. Photosensitive automatic porch light.*

pair. Almost any low-power NPN transistors should work well in this circuit. The exact operating parameters aren't critical. It would be a good idea to first breadboard the circuit to make sure it will work with the specific components you are using in your project, but it is doubtful that you're likely to run into any significant problems with this circuit.

Table 5-8 Suggested parts list for Project 32. Photosensitive automatic porch light.

IC1	555 timer
IC2	CD4011 quad NAND gate
Q1, Q2	NPN transistor (2N3904 or similar)
D1	diode (1N4002 or similar)
PC1	photovoltaic cell
K1	relay to suit load
F1	fuse to suit load
C1	150 µF 25 V electrolytic capacitor—see text
C2	0.01 µF capacitor
R1	680 kΩ ¼ W 5% resistor—see text
R2	5 kΩ potentiometer
R3	100 kΩ ¼ W 5% resistor

The output signal from the transistor amplifier stage is used to trigger the timer (IC1 and associated components) when the lighting level suddenly drops. The signal is ignored by the rest of the circuitry until the time delay period is over. After the timing period, if the detected lighting level is below a specific threshold level, IC2 will activate the relay (K1), turning on the light connected to socket SO1. If the detected light level is above this threshold value, IC2 deactivates the relay, and the controlled lighting (or other electrical) device will receive no power.

Select the value of fuse F1 to suit the intended load for your project. Don't use anything more than about a 2 or 2.5 A fuse in this circuit. This will permit a load of up to 240 to 300 W, which should be quite sufficient. For safety reasons, an appropriately valued fuse should always be used in any circuit carrying ac power.

The relay should also be selected so its switch contacts will carry more than the required load current. Diode D1 protects the delicate relay coil from possible back EMF spikes when it is switching states. Almost any standard silicon diode should work for this application.

Notice that ac power is carried only in the output portion of the circuit (from the relay switch contacts onward). The main body of the circuit (IC1 and its associated components) must be operated from its own dc supply voltage. Anything from +6 to +12 V can be used. You can either use battery power, or an ac-to-dc converter. In this particular application, The second option would be the most logical choice.

If suitable for your application, you might want to add a second SPST switch to cut off the dc supply voltage to the control circuit. In other applications, it might make more sense to operate the project continuously.

The sensitivity of the light detection circuit is set by potentiometer R2. Adjust this control for the circuit so the circuit responds to the desired light level. In calibrating the project, the time delay might be a little frustrating, but this is easy to overcome. Temporarily break the connection to pin 2 of IC1. This will effectively disable the time delay function, because the timer will never be triggered. In some cases, it might be a good idea to temporarily connect the trigger input (IC1, pin 2) to the V+ supply through a pull-up resistor, to avoid possible instability or random triggering while you are attempting to calibrate the circuit. Once you have adjusted R2 so the circuit responds to the desired light level, reconnect pin 2 of IC1 as shown in the schematic. Remember to remove the temporary pull-up resistor, if used.

In most applications it would be a good idea to use a screwdriver adjust trimpot for R2, but you can certainly use a front-panel, full-size potentiometer, if that suits your individual intended application.

Remote burned-out bulb indicator

Remote and automated light systems can be very useful and convenient, but there is one common problem with them. Generally the operator does not (or often cannot conveniently) check to make sure that the lighting is actually working. The control system could be working properly, but if one or more light bulbs are burned out, the entire system could be rendered entirely useless.

Many people with outdoor lighting systems in residential areas discover that some kids seem to enjoy stealing or breaking light bulbs for no discernible reason. Obviously, your remote or automated lighting system won't do much good if the bulbs are broken or burned out.

How often have you seen someone driving with only one working headlight? Unless someone tells them (usually a policeman, who might give a ticket along with the information), they don't know the light is burned out. Of course, in this example, there could be some very serious risk involved.

Believe it or not, a very similar problem was apparently one of the many contributing causes to the famous accident at the Three Mile Island nuclear power plant some years ago. The main problem was a stuck valve, but a burned-out indicator lamp concealed the problem and confused the issue until considerable damage had already been done.

If you use a remote or automated lighting system, you could develop a habit of periodically going out to the remote location to manually confirm that all the lights are, in fact, working. This project offers a more elegant and more reliable solution. An indicator lamp (or some other indicating device) is activated at the controlling station if the remote light bulb is burned out or missing.

The schematic diagram for this project is shown in Fig. 5-10. The parts list is given in Table 5-9. As you can see, this is not a particularly difficult or complex project.

This project is designed on the assumption that the lighting devices being monitored are powered by ac current. This circuit will not work with a dc powered system.

Looking at the circuitry, the primary of the transformer (T1) is placed in series with the remotely located lamp. As long as the lamp is working, the circuit is complete, and current flows through the transformer, holding the relay (K1) in is activated state. Diode D1 and capacitor C1 form a simple half-wave rectifier circuit feeding the relay's coil with a signal roughly resembling 12 Vdc. Diode D2 protects the relay coil from burning itself out due to back EMF during switching.

The indicator lamp (I1) is connected to the normally closed (NC) contacts of the relay. That is, when the relay is not activated, the indicator bulb will light up. When the relay is activated, the indicator bulb remains dark. If the remote lamp is burned out or broken, or if the bulb is removed, the circuit will be broken. No current will flow through the transformer, and the relay will be released, dropping into its normal, deactivated state, lighting up the indicator lamp.

For most applications, the indicator lamp (I1) should be a small, panel lamp. In some applications, you might want to substitute some other indicator device. Even an audible alert could be used, if that suits your needs the best.

Notice that the power connection for the indicator lamp (or other warning device) is made after the remote control switch. If the indicator lamp power was tapped off before the switch, it

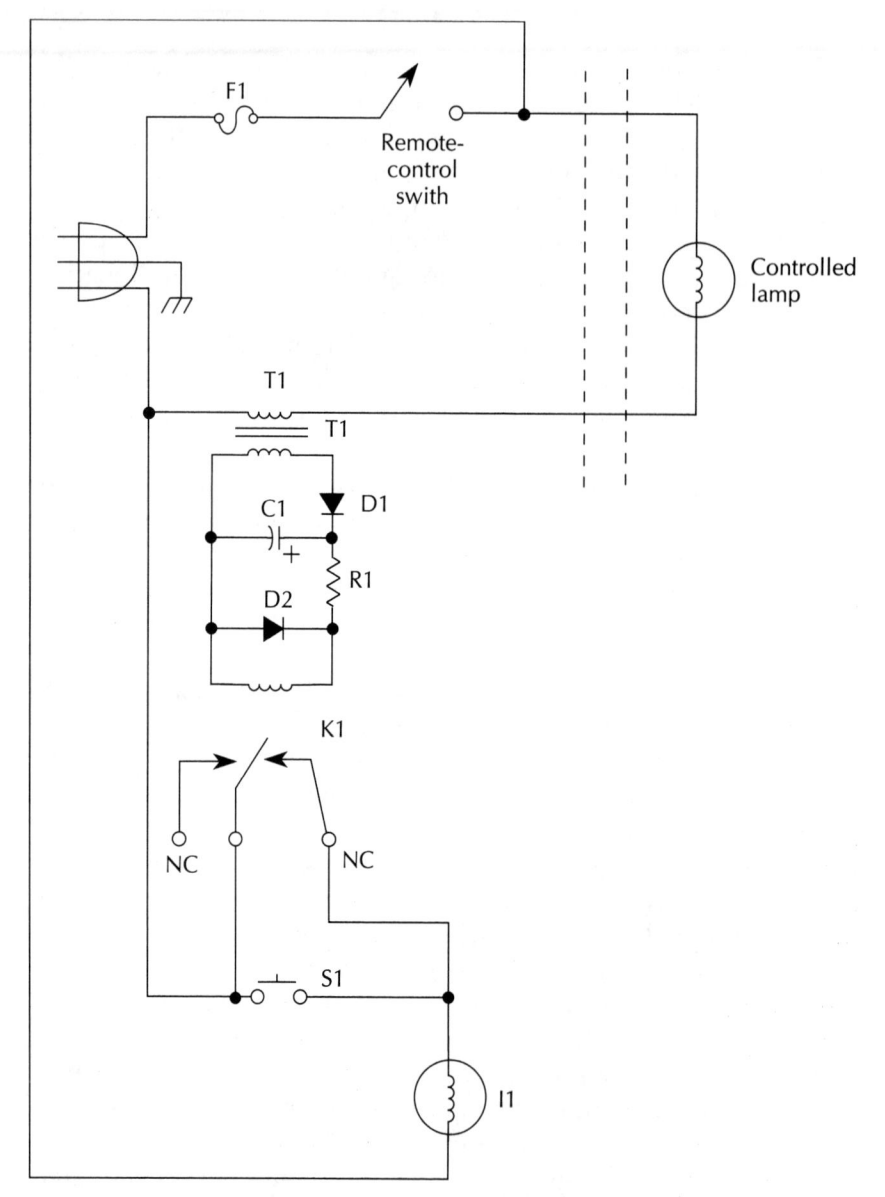

Fig. 5-10 *Project 33. Remote burned-out bulb indicator.*

would glow whenever the remote light was intentionally turned off. This would obviously serve no purpose. It would waste power (though, admittedly, not much), and would tend to shorten the lifespan of the indicator bulb.

An important part of this circuit is the normally open push-button switch (S1). This is what they left out of the design at

Table 5-9 Suggested parts list for Project 33. Remote burned-out bulb indicator.

D1, D2	diode (1N4002 or similar)
C1	100 µF electrolytic capacitor
T1	12.6 Vac power transformer
K1	12 V relay (contacts to suit load)
I1	Small indicator lamp—see text
S1	Normally open SPST push-button switch
R1	1 kΩ ½ W 5% resistor

Three Mile Island. Closing this switch can temporarily bypass the relay switch contacts, forcing the indicator lamp to light, regardless of the state of the remote lamp. This switch should be closed periodically to make sure the indicator lamp itself hasn't burned out. This can happen, and it would leave you no better off than if this monitoring circuit didn't exist at all—you won't know the condition of the remote lighting device.

The basic concepts of this project can be readily adapted for many other applications in remote control and automation systems of all types. Use your imagination.

6
Technological environmental risks

These are times of concern. Almost every week, the news tells you about of something new to worry about. Sometimes these reports are based on valid, even serious concerns. Other times they are based almost entirely on mistakes and misinterpretations, or even deliberate hoaxes. In almost all cases, the practical risk in the real world is often exaggerated in general news reports. There is a crying need for intelligent perspective in reporting ecology and health issues to the general public.

Modern technology is often presented as the chief villain in these environmental health risks. It seems most people fall into one extreme or the other. Either they are *technophobes* who consider most or all technology inherently bad and harmful, or they are *technophiles* who consider technology the answer to all of humanity's problems and consider any risks from technology insignificant or totally nonexistent.

The reality is probably somewhere in the middle. Some technology has a lot of real risks involved with it. Other types of technology are almost entirely safe (at least unless grossly misused). The media tends to exaggerate any technological risks because it sells newspapers and gets good ratings. Often, the highly publicized risk is actually relatively insignificant or unlikely, and the more real, but harder to identify, danger gets virtually ignored.

It is no longer possible or reasonable to write a book on technological theory without considering some of these ecological issues. The issues are inescapable in modern society. No one can afford to ignore them. Unfortunately, but inevitably, this will require a bit of a detour from the discussion of electronics theory, but the issues at hand are important enough to make such a detour worthwhile.

It is useful to present some general thoughts about such issues and how they are reported to and perceived by the general public. Then you can take a more informed look at the current concerns over the alleged health effects of electromagnetic fields, and other specific issues that relate to electronics technology.

The problem of proof versus disproof

There appear to be very few news reports on scientific issues that do not suffer from major errors of fact. Such errors are not due to deliberate or conscious bias. They are often the results of insufficient scientific training on the part of the reporters. Generally, a solid science background is not considered a requirement for science news reporting. Today, scientific and technological issues are vitally important. These issues cannot be adequately reported by someone who doesn't really understand them. No matter how good a journalist a person is, he or she can easily turn in a very distorted story, filled with errors and misinterpretations, if he or she does not understand which facts are valid and which are more or less irrelevant.

Scientific and technological issues are often complex, and most professionals in this area don't have very good communications skills. They can write reports that are understandable by their colleagues, but they all too often don't really know how to break down the relevant information so it is comprehensible by the general public. But, when a reporter without a science background interviews a scientist, and the reporter does not understand the underlying theories and principles, it is very easy for misconceptions to occur in the simplifying and rewording. Even using the scientist's exact words might lead to misinformation if quotes are taken out of context.

For example, scientific caution often leads to public paranoia over undefined risks. When a scientist says, for instance, that there is no 100% safe level of radioactivity, that is not necessarily grounds for concern. It certainly doesn't mean that all radioactivity is inherently bad. There is some scientific evidence that indicates that life would not be possible without some minimal amount of radioactivity. Even a level of zero would be unsafe in the sense the scientist means.

Radioactivity occurs when an atom emits subatomic particles and energy. When certain subatomic particles hit a chromosome, they can cause genetic damage. The more subatomic particles there are in a given area (that is the higher the level of radioactivity), the better the odds that a particle will hit a chromosome and do damage. As long as there is at least one subatomic particle, there is always some chance that it might make a lucky hit. The odds would be overwhelmingly against a hit under such circumstances, but it is not impossible. A scientist would not say the situation was completely free of risk, unless the chances of the negative event occurring are truly zero, without rounding off. If there is 1 chance in 1,000,000,000,000, it still counts as a risk to the scientist. So when a scientist says no level of radioactivity is completely safe, he or she just means the risk is never truly zero, but the general public can interpret the statement as meaning that any radioactivity at all is dangerous.

Another example is some reports that scientists had not ruled out the possibility of AIDS (acquired immune-deficiency syndrome) being contracted by kissing. When scientists were asked if it was possible to catch AIDS by kissing, they replied that such a possibility had not been conclusively disproven, even though there was no scientific evidence to suggest it was possible. The odds were very strongly against such a connection, but the scientists didn't want to claim to be all knowing. It is always extremely difficult to conclusively prove that there is a true 0% chance of contracting any disease by any given means. The specific connection has been neither proven nor disproven, and existing evidence suggests it is an unlikely connection.

Just because something has not been conclusively disproven does not mean it has been proven. Similarly, just because something has not been conclusively proven does not mean it has been disproven. It seems that many, if not most lay people, and even some scientists and technical professionals with excellent academic credentials, have great difficulty making this critical distinction. It is vital to understand this to intelligently assess any technological risk.

There is a critical difference between proof and evidence that is often ignored, even by professionals and experts. Inevitably this leads to foolish dogmatism and misinterpretation of facts. Proof is absolute and unquestionable. In the real world, true scientific proof is very, very rare, and perhaps even nonexistent. There is never any intelligent controversy over proof. Evidence, on the other hand, is subjective by definition. It requires interpretation. Two equally qualified experts might interpret the exact same piece of evidence in two very different, and perhaps even totally opposite, ways. To complicate matters further, they might not agree on just what is or isn't valid, meaningful evidence for the question at hand. One expert's conclusive evidence might be unconvincing, or even totally irrelevant to another expert with equal credentials.

An experiment that gives very impressive results, which might look like conclusive proof to the lay person (and some experts) might later be found (probably by other experts) to be seriously flawed in its procedures, perhaps in a very subtle and unexpected way, so the results are not truly meaningful at all. There is a tendency to say "such-and-such an experiment (or series of experiments) proved this-and-that conclusion." In almost all cases it would be more accurate to say that such-and-such an experiment (or series of experiments) *indicated* this-and-that conclusion."

Lay persons must always bear in mind that all experts in every field are human beings too, with the same sorts of failings, unconscious biases, and blind spots as the rest of humanity. Every expert is wrong in her or his field of expertise at least some of the time.

Any time there is a controversy on any scientific issue, it means that the experts disagree over the interpretation of the available evidence.

Curiously, and perhaps surprisingly, the lay person is more likely to be exposed to the radical, minority viewpoint in scientific controversies. An interested, intelligent lay person might even find it difficult to find nonprofessional materials presenting the majority viewpoint. This is because a radical, sensationalist theory will sell a lot more books, magazines or newspapers, but a similar publication covering a conservative, traditional theory will appear dry and dull to the lay person, who will probably leave it on the rack. So the conservative, mainstream scientific viewpoint is not as potentially profitable to publishers, so such lay-person books are less frequently written and published. The mainstream scientific viewpoint is usually covered in some detail in the technical and professional journals for the field in question. The wilder and the more sensational the theory, the better it will sell to the general public.

Notice that this discussion is not trying to prove that such theories are automatically wrong. The mainstream scientific community has been dead wrong many times in the past and will similarly err many times in the future. The point is, such sensationalist theories should not be accepted as gospel. The more media hoopla there is over any scientific idea, the more controversial it probably is, and the more strongly it goes against mainstream scientific thinking.

It is rarely, if ever, valid or reasonable to conclude that the experts who oppose the sensationalist theory have a vested interest, or are in the pay of those who do. If there were not good scientific reason to doubt the new theory, there would be no real controversy over it. The theory might ultimately turn out to be completely true. As long as the controversy exists, there is room for doubt, and it is unscientific and even foolish to insist "such-and-such is absolutely true" or "such-and-such is absolutely false."

Many authors, even those with strong academic credentials in their field, will write a sensationalist book more for monetary than scientific reasons. They might (perhaps not completely consciously) exaggerate the case to sell more books. Read any scientific literature oriented toward the lay person cautiously, and never let any one author (even the author of this book) fully convince you on any issue.

Often the title will give you a clue about the author's intent in writing the book. For example, one book on the subject of the alleged health risks of electromagnetic fields, which is discussed below, has the lurid, sensational and alarmist title *Currents of Death*. Such a wild, almost paranoid title should raise some doubt in the intelligent reader. In fairness to authors, realize that sometimes publishers change the author's title to help the book sell more copies. Check out the text before drawing conclusions. If a book has a wild title but cautious, well-reasoned arguments in the text, the title might not have been the author's intended title. But often, the text will be just as wild-eyed and paranoid as the title. The author might be sincere, but lacking perspective on the issue. Many fully qualified experts have an axe to grind that might get in the way of their scientific judgment.

In fairness to you, realize that the opinions of the author of this book are expressed before you leave the subject. You are not obligated to share the author's views. The views are given only so readers can compensate for any unintended bias in the discussion.

Electromagnetic-field detector

There has been a great deal of worry among the general public about the alleged health effects of electromagnetic fields, specifically ELF (extremely low frequency) fields. Although there is some scientific evidence for some concern about this problem, it has unquestionably been exaggerated.

ELF fields definitely don't present a critical health threat to human beings. If they did, the effects would have been noticed long ago. Humans have been living among increased ELF fields for over a century now, especially in the last 30 years or so. If there were a critical effect on health, there should be major indications of it by now. Population has increased enormously with the increased usage of electricity and the resulting ELF fields. (Note that a causal relation is not being suggested—the population explosion is not caused by the increased use of electricity.)

Nonetheless, even if the health risk is not critical, it might still be cause for concern. A health risk is a health risk, and the more you can avoid, the healthier you will be. The exact level of health risk (if any) from ELF fields has not been scientifically proven, although there is apparently at least some minor effect.

In looking at the scientific reports on such issues, you need to consider that when a scientist speaks of a significant health risk, she or he means something quite different from what the general public understands. For example, assume the incidence of a given disease among the general population is about 4.5%, but when a given factor is present, the incidence of that disease jumps to 7%. This is statistically significant, and not to be ignored, but it is hardly cause for alarm and panic. Except for a handful of extremists, most qualified scientists today consider the health risks from ELF fields to be "significant" in this sense, or negligible. It is an area that demands more research, and perhaps some simple, common-sense precautions, but serious fear seems unwarranted.

Take a moment to understand the terms of this controversy. Whenever current flows through a conductor, that conductor will be surrounded by an electromagnetic field. Electronic hobbyists understand this fact from their work with inductors. Strong, high-frequency electromagnetic fields can resonate with the inner structure of a cell, causing cells to overheat and damaging the cell. For example, most people have heard about the risk of long-term exposure to high-level X rays. An *X ray* is essentially an electromagnetic field with a very, very high frequency. In all cases, the field intensity has to be fairly strong to do damage.

Traditionally, it was assumed that low-frequency electromagnetic fields (ELF fields) couldn't present a health risk, because the wavelength is too long. Nothing on the cellular level is nearly large enough to resonate with a 60 Hz electromagnetic field. And it is true, that the overheating effects described above definitely do not occur with ELF fields.

However, recently evidence has been found that suggests that cells can, under some circumstances, react to, and even be damaged by ELF fields in different ways. No one currently knows just what mechanisms are involved here or how the damage is done. But remember that the overall effect is less than the effects of high-frequency electromagnetic fields—that is why the effects of ELF fields weren't discovered first.

On the other side of the coin, there is some evidence (hardly conclusive yet) that suggests ELF fields can cause their effects at surprisingly low intensities.

Should ELF fields be considered unsafe at any intensity? Only if you want to worry yourself to death. You're never going to eliminate all exposure to ELF fields.

First, be aware that everyone is always surrounded by electromagnetic fields. These fields occur naturally, and there is some fairly strong scientific evidence indicating that if natural electromagnetic fields did not exist, life itself could not be sustained. In many respects, the earth itself is a giant magnet.

Any reasonable discussion of the health aspects of electromagnetic fields must begin with an understanding of the natural electromagnetic background everyone is continuously exposed to.

Because electromagnetic fields are completely natural phenomenon, any risks associated with them must be of quantity, not of kind. Despite some of the more paranoid and sensational of some of the writings on this subject, electromagnetic fields are not inherently harmful.

All electromagnetic fields are force fields. That is, they carry energy and can produce an action at a distance. For example, a permanent magnet can move a small metallic object some distance away. The farther from the source, the weaker the electromagnetic field gets. The field strength drops off rather quickly, following a logarithmic rather than a linear pattern.

In a sense, the planet Earth is a gigantic bar magnet. The

north and south poles of magnets are named for their electromagnetic similarity to the earth's North Pole and South Pole. The true magnetic poles of the earth are not entirely stationary. They move around slightly, and are usually not located precisely at the true geographic poles.

There is some scientific evidence that from time to time in the past, the electromagnetic field of the earth has reversed polarity. That is, the north magnetic pole became the south magnetic pole, and vice versa. Some scientists hypothesize that such large-scale magnetic field reversals were responsible for the periodic die outs of species that have mysteriously occurred in the earth's past. This theory is quite intriguing, but it is very, very controversial, and cannot be accepted as proven fact. Many, perhaps most scientists today do not believe any such magnetic pole reversals have ever occurred, or even that they could occur. The existing evidence, although somewhat impressive, is still highly questionable.

Many (not all) authors who believe in the harmful effects of electromagnetic fields tend to accept such controversial theories as given facts. This places their hypotheses on shaky ground right from the start.

The core of the earth is molten rock, very heavy in iron. The spinning core of the earth creates a dipole magnetic field with a magnetic north pole and south pole. As with any permanent magnet, force lines extend from pole to pole, as shown in Fig. 6-1. This illustration is not accurate, however. It shows what the magnetic fields of the earth would look like if it was alone in the universe. But the earth is far from the only electromagnetic object in the universe. It's electromagnetic fields are inevitably acted upon by other cosmic objects. The primary influence is the sun, because it is so large and relatively close. For your purposes, assume that only the electromagnetic interference from the sun is of significance. The effects from other planets, the moon, and nearby stars are real, but far, far weaker and more subtle than the effects from the sun. They don't change the overall picture significantly.

The sun constantly emits a force known as the *solar wind*. This *wind* is not a true wind, in the way you normally think of it here on earth. An ordinary wind is a movement of air, and there is no air in space. The solar wind is a flow of high-energy atomic particles emitted from the surface of the sun. In many respects, a solar wind acts rather like an ordinary earth wind, so the name is appropriate, as long as you don't take it too literally.

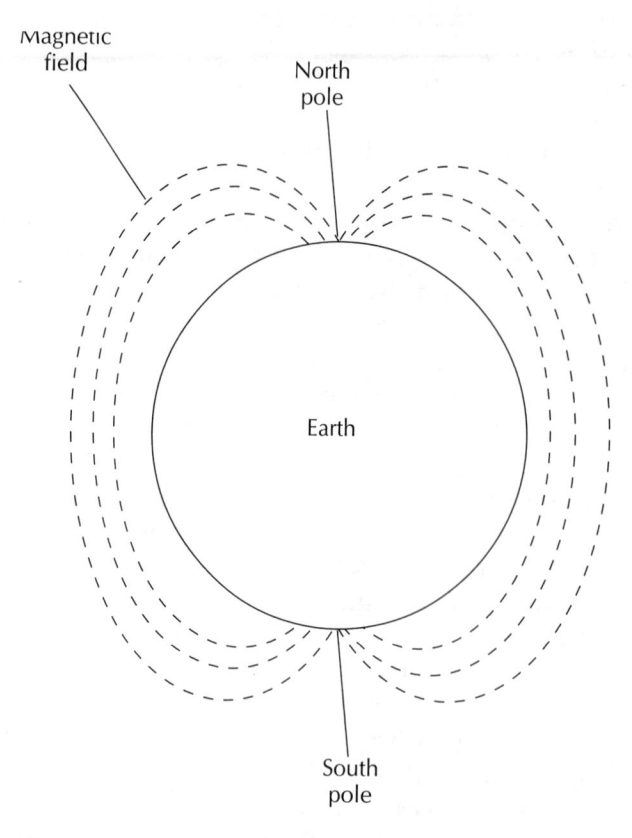

Fig. 6-1 Conceptually, the earth acts like a gigantic permanent magnet.

The solar wind contains particles with very high energy moving at high speeds. Some of these high-energy particles are of the ionizing type, and others are non-ionizing. There is still plenty of energy left in the solar wind by the time the particles travel the distance from the sun to the earth. The solar wind therefore interacts with the earth's natural electromagnetic fields. On the side facing the sun, the solar-wind particles push against the earth's magnetic fields, compressing them. Meanwhile, the fields on the far side of the planet are "blown" outward by the solar wind to form a long *magnetotail*.

The collision of the solar-wind particles with the earth's magnetic fields creates a *bow-shock region* in which these forces interact. Two special areas within this bow-shock region are known as the *Van Allen belts*. Some of the solar wind's high-energy particles are trapped within these belts, where they constantly spiral between the north and south ends of the *ducts*, as shown in Fig. 6-2.

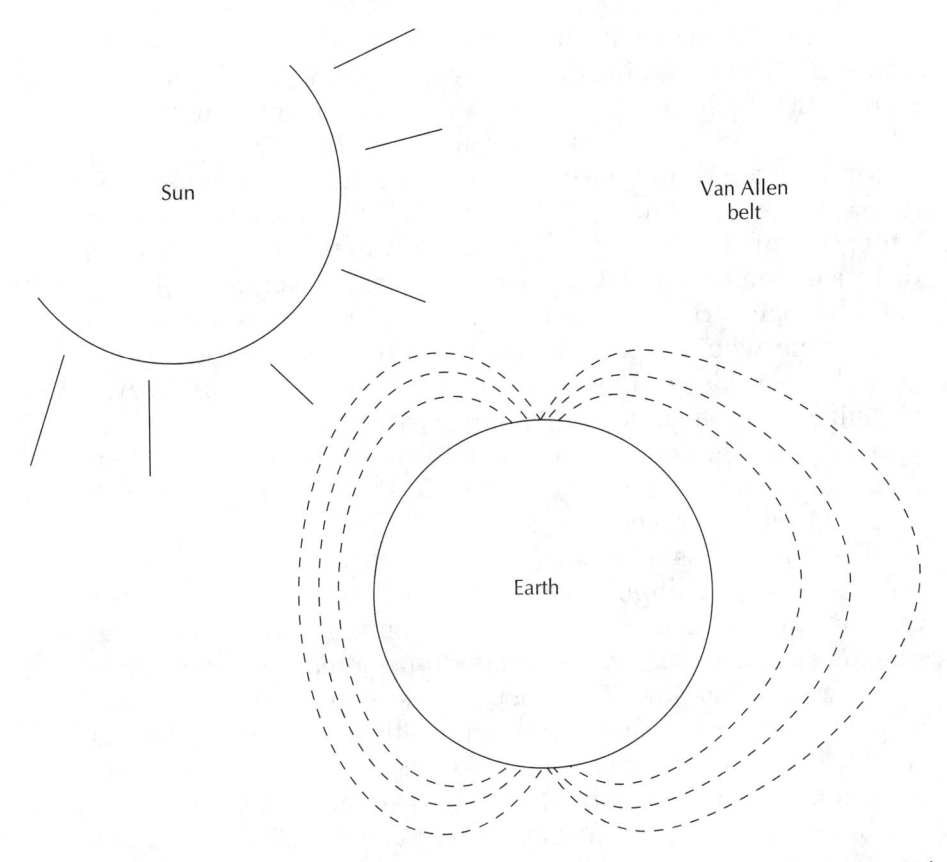

Fig. 6-2 *Some of the high-energy particles from the solar wind are trapped within the Van Allen belts and constantly spiral between the north and south ends of the ducts.*

The *magnetosphere* (the distorted magnetic fields surrounding the earth) shields the planet from much of the sun's radiation, especially the potentially harmful ionizing rays. If the magnetosphere were destroyed or removed, all life on earth would cease to exist. Space flights beyond the earth's magnetosphere must be of limited duration to prevent harmful effects to the astronauts caused by exposure to the sun's powerful radiation without this natural shield. (Theoretically, an artificial magnetic shield could be designed and incorporated into future spacecraft, but such technology does not yet exist.) Such space flight missions beyond the earth's magnetosphere must also be timed carefully so they occur during quiet periods in the earth's cycle. During a solar storm, the ionizing radiation emitted from the sun is much greater.

The earth rotates within the magnetosphere, which remains stationary. The same side of the magnetosphere always faces the sun. As the earth rotates under the magnetosphere, a given spot on the planet's surface will experience a daily pattern of up and down fluctuations in the strength of the natural electromagnetic field. At certain times of day, the electromagnetic field is stronger, and at other times it is weaker. Some scientists believe these electromagnetic level fluctuations might help explain the daily biological rhythms that occur in many species. For example, people who are placed in caves or enclosed buildings for extended periods with no way to tell time will still tend to synchronize their sleep/waking patterns with the earth's night/day cycle, at least approximately. Actually, there seems to be a tendency to act as if the day were a little shorter than the earth's actual 24-rotation speed.

This theory has a lot to recommend it. However, before ascribing too much importance to this still controversial theory, remind yourself that there are almost certainly other physical phenomena that have daily fluctuation patterns synchronized to the earth's rotation. The fluctuations in the natural electromagnetic field might have nothing at all to do with biological rhythms, despite the suggested evidence. It seems like they do have at least some effect, but this is far from proven to date.

Of course, the current controversy about ELF fields is not concerned with the relatively low, natural electromagnetic fields present all over the earth, but the relatively strong artificial ELF fields generated by the electrical equipment. Most ac power (in the United States) uses a frequency of 60 Hz, so the ELF fields you are most exposed to have this frequency.

The greatest risk (if any) comes from high-power overhead power cables and heavy industrial equipment. More controversial are the effects of the significantly lower ELF fields associated with ordinary home and office appliances.

Not all electrical devices produce any ELF fields at all (except for the power cord and the wiring in the walls—using the device in question has little practical effect on these fields). Generally, the concern is only over inductive loads. Remember, any inductor (coil, transformer, electrical motor, etc.) operates by creating electromagnetic fields as current flows through the conductor in the inductor. Unless the inductor is very large, this induced electromagnetic field won't be very large. It will also be restricted by the amount of electrical power consumed by the device—nothing can put out more energy than it takes in.

More important, the strength of the electromagnetic field drops off very rapidly as distance increases. For example, measuring the electromagnetic field from the front of a typical computer monitor gave a rather high reading of a little over 75 milligauss right at the screen. But moving just six inches away from the screen reduced the reading to under 20 milligauss. At one foot away, an ELF field of less than 5 milligauss was detected. Beyond two feet, the field intensity was less than 1 milligauss. This suggests the most practical and reasonable precaution to take if you are concerned by ELF fields—move a little farther from the ELF generating device in question. A very small increase in distance makes a big difference.

Incidentally, computer monitors are often singled out as particularly important in anti-ELF writings. A typical computer monitor does not put out more ELF energy than a typical television set. This is because a computer monitor and a television set are essentially the same thing electronically, except a television set has some tuner circuitry lacking in the computer monitor.

The only significant difference (for the purposes of the present discussion) is distance. A computer operator will usually sit much closer to the monitor than someone watching TV. The solution, move the monitor a little farther away. About 2 to 2½ feet should reduce the ELF fields to insignificant levels. Also, sitting closer to the screen would tend to lead to increased eyestrain and fatigue. It is interesting to note that the most common supposed effects of ELF radiation are headaches and fatigue, which could be symptoms of simple eyestrain. In either case, moving back a little from the screen should help the problem.

It should also be mentioned that a typical computer monitor or television set tends to emit more ELF radiation from the rear than from the screen. This is something to consider in an office with a large number of computer stations. Also remember that ELF fields can move easily through most standard building materials, so if the monitor or television set is against a wall, it will send its ELF fields into the next room on the other side of the wall. Again, this won't be a problem for someone in the other room, unless they are very close to the wall.

It also seems that the harmful effects of ELF fields, if they exist at all, are cumulative. Short-term exposure, even to relatively large ELF fields, doesn't appear to be of much importance. The alleged problems seem to come from long-term exposure. Therefore, even though an electric razor or hair dryer puts out fairly strong ELF fields, and is used very close to the body, it probably isn't doing much harm, because it is only used for a relatively short time.

If you are concerned about ELF fields, you probably would like some way to know when you are being exposed to a significant amount of ELF radiation. Electromagnetic meters are rather expensive, complex, and sophisticated devices, and probably outside the reach of most consumers. However, the circuit shown in Fig. 6-3 does a reasonably good job of giving a simple yes/no response. If a significant amount of ELF radiation is detected, a red LED will light up. Otherwise the green LED will be lit.

A suitable parts list for this project is given in Table 6-1. Nothing is too critical here, especially because no one really knows what the critical intensity is for ELF fields, so if the circuit is a little off in its measurements, this won't matter too much.

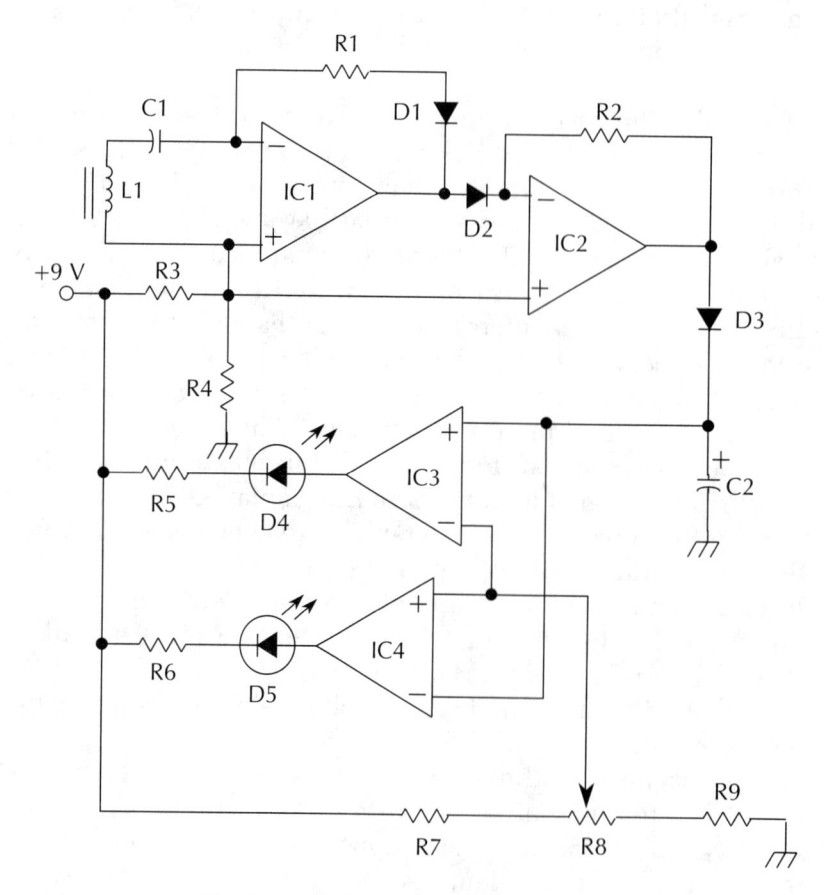

Fig. 6-3 Project 34. ELF field detector.

Table 6-1 Suggested parts list for Project 34. ELF field detector.

IC1, IC2	low-noise op amp—see text
IC3, IC4	op amp or comparator—see text
D1–D3	small signal diode
D4	green LED
D5	red LED
L1	telephone pick-up coil—see text
C1	0.33 µF capacitor
C2	4.7 µF 25 V electrolytic capacitor
R1, R2	1 MΩ ¼ W 5% resistor
R3	12 kΩ ¼ W 5% resistor
R4	8.2 kΩ ¼ W 5% resistor
R5, R6, R7, R9	1 kΩ ¼ W 5% resistor
R8	10 kΩ trimpot

The input device of this circuit is L1, a pick-up coil normally used for recording telephone conversations. It is a small microphone enclosed in a suction cup housing, meant to be adhered to the telephone handset. Of course, you are not using it for its intended purpose here. You can leave it in the suction cup housing (it's hard to get it out without damaging the pick-up coil anyway). This will actually be handy for monitoring the ELF radiation directly at a source device.

These telephone pick-up coils are inexpensive and widely available. Radio Shack sells them for just a couple dollars apiece.

The magnetic signal detected by the pick-up coil (L1) will be quite weak, so IC1 and IC2 should be low-noise op amps. BiFET (bipolar field-effect transistor) op amps would be a good choice. IC3 and IC4 are much less critical. These op amps are used as simple voltage comparators. You could use half a quad LM339 quad comparator IC for IC3 and IC4 if you choose, but remember the LM339 cannot be used for IC1 and IC2 in this circuit.

Be sure to make the appropriate power supply connections to all op amp ICs in the circuit. They are not shown in the schematic to avoid cluttering it, and also to allow for individual variations. Some readers might use four single op amp chips, and others might use dual or quad op amp ICs. Also, bear in mind that some op amp devices will work on a single polarity power supply (V+ and GND), and others require a dual polarity power supply (V+ and V−, both referenced to ground). Check the manufacturer's specification sheet for the specific device(s) you are using in your project.

Almost any standard signal diodes should work well for D1, D2, and D3. For maximum sensitivity, you can use germanium diodes, which have a lower internal voltage drop than silicon diodes. The difference will be very significant, but some people might disagree and be concerned about lower-intensity ELF fields.

LED D4 should be green, and LED D5 should be red. The different colors will make it a lot easier to read the detected electromagnetic condition at a glance. At the very least, mark the LEDs clearly. When D4 is lit the detected ELF energy is within the acceptable or normal range. When D4 goes dark, and the red LED (D5) glows, it indicates that a higher than normal level of ELF energy is being detected, and you should respond in whatever way you consider appropriate.

The supply voltage for this circuit can be anything from +9 to +15 V. If you use a +12 V power supply, increase the values of resistors R8 and R9 to be from 1.5 kΩ to 2.2 kΩ each. If the power supply voltage is +15 V, the values of R8 and R9 should be 1.8 kΩ to 2.7 kΩ each.

To calibrate this electromagnetic field detector, take the unit to an area relatively free of ELF radiation. An open field would be ideal but not absolutely necessary. A room with no inductive electrical equipment running will do. This will be the base radiation level everything monitored by the circuit will be compared to. Adjust potentiometer R8 so the green LED (D4) is fully on and LED (D5) is fully off. Now, slowly adjust the potentiometer until D4 just goes off and D5 comes on, then back off slightly—just enough so that the green LED (D4) is securely lit. That's all there is to the calibration procedure. Now, whenever the pick-up coil sensor (L1) detects an ELF field higher than this reference level, it will turn on the red LED (D5).

Don't use a front-panel control for R6. Use a good trimpot that will hold its set position well. You could use a ten-turn trimpot, but that would be overkill in this application. The calibration of this circuit just isn't all that exact, so a super-accurate trimpot won't give much practical advantage.

This circuit responds very quickly to any increase in the detected ELF intensity, but for some reason, it is a little slower to respond when the detected ELF energy drops back down into the acceptable range. But even here, the delay is just a second or two, so the project still works pretty well for most practical purposes.

Radioactivity monitor

Perhaps the greatest of all the technologically based fears of today's society is the fear of radioactivity. The general public's attitudes about radiation very definitely border on paranoia, and often go right over the edge. A lot of nonsense has been published on this subject.

The paragraph is not intended to suggest that there isn't legitimate grounds for serious concern in this area, and perhaps, even some fear under certain circumstances. No one would claim there is never any danger associated with radioactivity. The point is, the specific fears of the general public are often greatly exaggerated, and more important, often misplaced. Lay persons all too often get very worked up over relatively harmless irrelevancies, demanding unnecessary, redundant, and expensive laws and regulations, and the true risks go ignored as often as not.

The sad fact is, the loudest protesters (who presumedly mean well) often don't really know what they are talking about. Even when they have valid scientific support for some of their views, they often gabble it into meaninglessness. Some years ago, there was a radio program about a major antinuclear power protest. It was perfectly clear that the producers sided very heavily with the protesters. (There was no claim of objective news reporting made.) Several of the protesters were interviewed on this program. It was fascinating that, given the obvious slant of the program's producers, every one of the protesters said something along the lines of "No, I haven't the faintest idea of how nuclear power works, but I know it is dangerous and bad." In other words, each of the interviewed protesters proudly declared that his or her opinion was basically an uninformed one. Any valid points they might have were deeply buried in uninformed and probably incorrect information.

This book is certainly not the place for a lengthy discussion of the pros and cons of nuclear power, but a few brief points are relevant. The basic thrust of these comments is applicable to almost any technological threat.

First, nothing is risk free. A number of people have said things like, "If nuclear power endangers even one person, then it's not worth it." A fine sentiment, but in the real world, it is ridiculous. Any kind of power source, even a simple campfire, involves some sort of environmental risk. The best designed coal- and oil-fired power plants inevitably dump tons of deadly poisons into the air, even when they are working properly. Even solar power is not truly 100% pollution free as is often claimed. Operating solar cells causes no pollution, but the manufacture of the solar cells themselves creates some of the worst pollutants known. Although solar power can take much of the load of power consumption, it is very doubtful that it will ever be able to support heavy industry, so society will probably never be able to go to full solar power, without some supplementary power sources.

And probably the riskiest option of them all would be to have insufficient power to meet society's needs, especially industrial needs. If you think there is an unemployment problem now, consider what would happen if all factories closed down because there wasn't enough power available to keep them running.

Society must live with the fact that any possible choice contains some inherent risks. You have to carefully decide which options present the most acceptable risks and the best benefits. The person who says, "No risk is acceptable!" is just burying his or her head in the sand. Although there are realistic concerns about health hazards from radiation, statistically nuclear power is only a tiny, insignificant element in the whole picture. Except for people involved in certain occupations, the major radiation risks faced by most people are natural in origin, not man-made.

Before going further, you need to define terms. Very simply, *radiation* is energy that is emitted from some source, and it can travel over distance. Many types of energy fit this definition. Mechanical energy is an example of a type of energy that does not qualify as radiation. Not all radiation is harmful. Some of it is very beneficial or even essential to life. Light and heat are two good examples of types of radiation no one would want to do without. Of course, any type of energy can be dangerous if its intensity is high enough. For example, heating the air in your immediate environment to about 70°F is good and healthy, but to heat it to 2300°F would be deadly. It's a case of too much of a good thing.

When most people speak of the dangers of radiation, they actually are referring to a specific type of radiation known as *ionizing radiation* or *radioactivity*. Again, to oversimplify somewhat, radioactivity is energy caused by the emission of subatomic particles or waves. There are several different types of radioactivity, which are discussed below. Not all types of radioactivity are particularly dangerous.

Radioactivity is called ionizing radiation, because it can knock electrons off any atoms in its path, leaving them with an electrical charge. An electrically charged atom is called an *ion*. Ionizing radiation is simply radiation that creates ions.

Not all energy involving subatomic particles is considered radioactivity or even radiation. A familiar example for readers of this book would be electricity, which involves the movement of specific subatomic particles (electrons) through wires or other conductive materials. The path travelled by these electrons is more or less restricted (a specific current path), and very few of the electrons are emitted (or radiated) into the surrounding atmosphere under most normal conditions.

You have probably heard that scientists agree that no level of radioactivity is 100% safe. Unfortunately, scientists don't mean this statement quite the way the general public usually interprets it. Radioactivity can do cellular damage whenever one of the emitted particles strikes a chromosome, altering its structure in some way. In most case this collision will simply kill the cell, and because millions of cells die on their own, and are replaced every day, this generally isn't very important in itself. But occasionally, the radioactive particle might hit just right to change the genetic code in the chromosome. This can cause a mutation or a cancerous condition.

The higher the level of radioactivity the body is exposed to, the more particles will be passed through the body, and the more likely it will be that a lucky (or unlucky) hit will occur. As long as there is at least one radioactive particle in the area, there is some finite chance of this happening. This is why no level of radioactivity can ever be considered 100% safe with true scientific accuracy.

Some scientists theorize that even zero radioactivity would not be a safe level, because there is some (highly controversial) evidence that minor levels of certain types of radioactivity might be essential to the maintenance of life, even though that same level of radioactivity might occasionally damage a specific individual.

Remember that you can't possibly avoid all radioactivity, no matter how much it scares you. Everything on earth is radioactive to some extent, even human bodies. All matter gives off (radiates) subatomic particles. For most materials, these particles are emitted very infrequently. Highly radioactive materials, such as uranium and plutonium give off very large numbers of particles in a short period of time. For example, in the time that a piece of lead gives off ten subatomic particles (any one of which could conceivably do cellular damage if it hits just right), a piece of uranium of the same mass might emit several thousand, or even millions of particles—increasing the odds of a harmful hit considerably.

The natural background radioactivity level varies in different parts of the earth. Mountainous regions tend to have more natural radioactivity than flat lands.

Coal is usually dug out of mountains, so it is radioactive. Some types of coal are much more radioactive than others, but all are radioactive to some degree. A pound of coal (of any type) doesn't emit very many particles, but a typical coal-fired power plant burns tons of coal every day. Most people are not aware that it is a well-established scientific fact that a properly functioning coal-fired power plant emits 80% to 500% more radioactivity into the atmosphere than a comparable nuclear power plant. (The wide range is due to variations in the types of coal that might be used as fuel.)

So, radioactivity is always all around. It is part of the nature of things. The higher the intensity of radioactivity in your immediate environment, the greater the risk it will present to you. And there is simply no way to define a magic cutoff point that eliminates all risk. It is a game of odds, no matter how you play it. For most of the general population, the risk from radioactivity is statistically much less than the risk of an automobile accident. You must keep the risk in perspective.

But some people are exposed to unacceptably high risk from radioactivity. People in occupations involving the handling of highly radioactive materials are obvious examples. Another good example would be an X-ray technician. (X rays are a form of radioactivity.)

The overall health risk of radioactivity is cumulative. Having one X ray taken doesn't increase your statistical risk noticeably, but if you have a couple hundred X rays taken in a year's time, there would be reason for some concern. (But no reason to panic. There is no guarantee that any harm will be done—just the odds are increased.)

A lot depends on where you live. If the natural background radioactivity in your area is low, you can accept somewhat higher levels of artificial (man-made) radioactivity.

One of the most serious radioactivity risks faced by the general public is radon gas. (Even here, statistically the risk is very low for most people.) It is important to realize that radon is a very natural phenomenon. Certain materials in the earth tend to emit radon gas (which is fairly high in radioactivity). Many homes and public buildings were erected using these materials before the potential risk was recognized. These problems usually occur in brick or stone buildings, rather than in wood buildings or modern steel structures. Not all brick or stone buildings are dangerous. Most are perfectly safe, and there is no practical radon gas threat to the occupants. But if there are any problems with radon gas, it will most likely occur in a brick or stone building. Radon gas problems can occur from other sources, but they are less common

In most areas, the local power companies or a governmental task force will check your home or office (or whatever) for abnormal levels of radon gas upon request. Sometimes there is a nominal fee for such tests, and in other areas it is free.

There are several different types of radioactivity, and they are not all equally risky. The three primary types of radioactivity are named for the types of particles they emit—alpha, beta, or gamma radiation.

Alpha radiation is composed of positively charged particles. Loosely, you could consider an alpha particle to be a proton. This isn't strictly accurate, but it is close enough for this discussion. On the subatomic level, an alpha particle is a pretty massive object. When it is emitted from a radioactive atom, an alpha particle will have a velocity of about 5% the speed of light. Five percent might not sound like much, but because the speed of light is 186,000 miles per second, the relatively slow alpha particle is moving at about 9300 miles per second—still a pretty good clip.

When an alpha particle collides with another atom in its path, it will knock one or more electrons off the atom it hit. This creates negative and positive ions. Because the alpha particle is fairly large (on the atomic scale), such collisions occur rather frequently. Some of the alpha particle energy will be used up in each collision. That is, each time it bumps into another atom and creates ions, it slows down by a significant amount. After enough collisions, it will have no more velocity than a typical air molecule. It will become part of the atmosphere, basically as a positive helium ion. Any free electrons (negative ions) in the area will combine with the former alpha particle to turn it into an ordinary helium molecule.

Considering the relatively large mass, and the high velocity of the alpha particle, you might well assume that alpha radiation would be highly dangerous. Actually, you'd be wrong in that assumption.

Alpha particles can easily be blocked by shielding. What do you need to effectively shield against alpha particles? Oh, a few sheets of ordinary newspaper should do the trick. Even in open air, an alpha particle emitted from a radioactive substance will typically only travel a few inches before neutralizing itself.

Alpha radiation is probably not worth worrying about. It presents virtually no practical risk. If your body is hit by alpha particles, they will be almost completely blocked by your clothing. Even if they hit your naked skin, they will be neutralized by collisions with the atoms in the uppermost layers of your skin. Skin cells on the outermost layer are continuously dying off and being replaced anyway, so any damage done by the alpha particles will be almost irrelevant.

In *beta* radiation, the particles emitted from the radioactive nucleus are much, much smaller than alpha particles. A beta particle is about 1/7500 as massive as an alpha particle. Beta particles resemble ordinary electrons in mass and electrical charge, but they are emitted from the nucleus. They are not taken from the orbiting electrons surrounding the atom.

Because beta particles are considerably smaller than alpha particles, they will experience fewer atomic collisions when passing through the same material. The tiny beta particle is more likely to go through the space between adjacent atoms. However, atomic collisions are still not infrequent. As with alpha radiation, when a beta particle strikes an atom, it knocks off one or more electrons, causing ionization. Some of the particle energy will be used up in each collision, and eventually it will more or less come to a stop, and act like an ordinary free electron, or negative ion.

A beta particle can typically travel several hundred times further than a comparable alpha particle, but it can still easily be shielded. A quarter inch thick sheet of Lucite will effectively shield against beta radiation.

If you are exposed to beta radiation, it will probably kill off a few more cells than alpha radiation of the same intensity, but you are very unlikely to notice any effects to your overall health, unless the intensity is extremely high. Again, beta radiation does not present much of a practical health threat.

Alpha particles have a positive charge, and beta particles have a negative charge. Either can be deflected by an appropriate magnetic field.

By this time you might suspect that the discussion is leading up to a claim that health concerns about radioactivity are completely unfounded. No, that is not true. You've still got gamma radiation to consider, and that one's the big one.

Unlike alpha radiation or beta radiation, *gamma* radiation is not composed of particles, at least not in the ordinary sense. Like light, it can sometimes act like a particle, and it can sometimes act like an energy wave. You could visualize alpha radiation and beta radiation as a stream of tiny ping-pong balls being shot out of a miniature gun, but this metaphoric image doesn't work for gamma radiation. It is much closer to radio waves, microwaves, and especially X rays. In fact, what are called X rays are actually a form of gamma radiation with a fairly low frequency. Gamma rays from radioactive sources have a very high frequency, and are extremely penetrating. They can travel tremendous distances through open air, and they can pass through almost anything. To adequately shield against gamma radiation, you need several inches of lead, or several feet of solid concrete.

If a living organism is exposed even to relatively low intensities of gamma radiation, massive cellular damage could occur. Obviously, this is the type of radioactivity to be concerned about. It can be very, very dangerous and even deadly.

However, you must maintain some perspective. You are all constantly being bombarded by low-level gamma radiation. All matter on earth is radioactive to some extent, although most ordinary materials emit only negligible amounts of gamma radiation under normal circumstances. The entire planet is also being continuously bathed in gamma radiation from the sun, and other nearby stars. This often called *cosmic* radiation. All types of radioactivity can travel incredible distances through space, where there are relatively few atoms in the path to create paths to slow down the particles or rays. Most cosmic radiation is safely absorbed by the atoms of the earth's upper atmosphere, but some (especially gamma rays) does get through to the surface. Some gamma rays pass right through the entire planet, and head out into open space from the opposite side.

This background gamma radiation is impossible to avoid, and actually is of little practical concern. Occasionally it might harm an individual, as a spontaneous mutation of one or more critical cells, but this event is relatively rare, or life itself would not be able to exist on earth. Although everyone is constantly exposed to low-level gamma radiation from natural sources, most never experience any ill effects from it.

Naturally, the higher the intensity of the gamma radiation you are exposed to, the greater the risk of serious cellular damage. This is why it is risky to work with highly radioactive materials, such as uranium and plutonium, which fortunately exist in only relatively small quantities in the surface of the planet. But remember, these radioactive substances are just as natural as carbon or oxygen.

So-called man-made radiation isn't really artificially created, so much as it is artificially concentrated and/or released.

One of the scariest things about radioactivity is that there is no way you can detect it directly, using any of your senses. You can't see it, hear it, smell it, taste it, or touch it. Some sort of equipment is required to detect and measure radioactivity, which finally brings you to the final project in this book—a crude, but effective radioactivity monitor circuit.

Usually radioactivity is monitored with a device called a *Geiger counter*, that causes a circuit to produce an audible click each time it detects a radioactive particle. The more clicks you hear, the higher the intensity of the radioactivity in the monitored area.

The input sensor for a Geiger counter is a special type of tube called the *Geiger-Müller* tube, named after its inventors. In the past, Geiger-Müller tubes were large, expensive, and relatively fragile, so they weren't practical for use in hobbyist projects. Recently, the technology has improved sufficiently to permit the manufacture of relatively small and not-too-expensive Geiger-Müller tubes. These devices are still considerably more expensive than any standard electronic components. A new, inexpensive Geiger-Müller tube can be purchased for as little as $60. Still pretty steep, but nothing like the old prices of $200, $300, or more.

Your best source for a Geiger would be an electronic surplus store, especially a military surplus outlet. You will also get the best possible prices from a surplus dealer.

This radioactivity monitor project is not intended as a full Geiger counter. It basically just detects the presence of abnormal levels of radioactivity, but doesn't attempt to measure the actual intensity. That would make the project too involved and expensive to be practical.

The schematic diagram for the radioactivity monitor circuit is shown in Fig. 6-4. The complete parts list for this project appears as Table 6-2. It is recommended that you not experiment with alternate component values in this project.

Radioactivity monitor

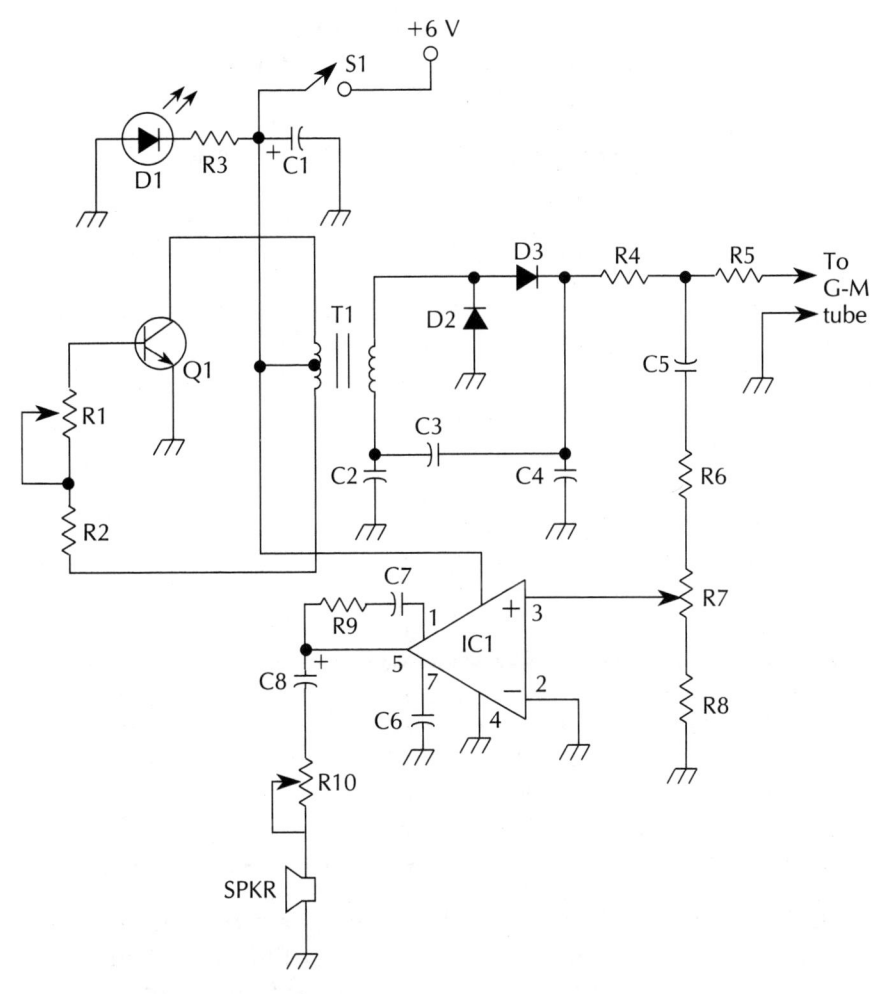

Fig. 6-4 *Project 35. Radioactivity monitor.*

Be very careful in building and using this project. Make sure everything is fully insulated and shielded. The project includes some very high-voltage circuitry, which could be a dangerous shock or fire hazard if you are careless. The high voltage is needed to drive the Geiger-Müller tube. Power transformer T1 is used to create the necessary high voltage. Notice that this transformer is being used backwards, with the supply voltage being fed into the secondary (smaller) coil, and the output voltage taken off the primary (larger) coil. In other words, you are using it as a step-up transformer in this circuit. Notice that this transformer must include a center tap in its nominal secondary coil.

Table 6-2 Suggested parts list for Project 35. Radioactivity monitor.

IC1	LM386 audio amplifier (or similar)
Q1	NPN power transistor (2N3055 or similar)
D1	LED
D2, D3	diode 1A, 1000 PIV (peak inverse voltage)
T1	24 Vac 450 mA power transformer with center tap
S1	SPST switch
C1	1000 µF 25 V electrolytic capacitor
C2, C3, C4	0.0047 µF 1000 V capacitor
C5	47 µF 1000 V capacitor
C6	0.033 µF capacitor
C7	100 µF 25 V electrolytic capacitor
C8	220 µF 25 V electrolytic capacitor
SPKR	small 8 Ω speaker
R1	1 kΩ ten-turn trimpot
R2	8.2 kΩ ½ W 5% resistor
R3	330 Ω ¼ W 5% resistor
R4, R5	470 kΩ ¼ W 5% resistor
R6, R8	1 kΩ ½ W 5% resistor
R7	10 kΩ trimpot
R9	10 kΩ ¼ W 5% resistor
R10	500 Ω potentiometer
GM tube	Geiger-Müller Tube—see text

Each time the Geiger-Müller tube senses a radioactive particle, it will generate a pulse, which is amplified by IC1 and its associated components. Actually, you could use almost any standard audio amplifier circuit for this portion of the project.

To calibrate this project, first make sure the power is completely disconnected from the circuit. Remember, you are dealing with a high-voltage circuit. Please, don't take any foolish chances. Connect a voltmeter probe at the junction between resistors R4 and R5, and connect the voltmeter ground clip to a convenient circuit ground point. The voltmeter must be able to read at least 600 V, or you risk damaging your meter. Leave the Geiger-Müller tube disconnected at this point.

Once everything is hooked up correctly, carefully apply power (+6 V) to the circuit and close power switch S1. The LED (D1) should light up. This LED is a simple power indicator, and you shouldn't have any problems here, unless you happen to have a bad LED.

Carefully adjust trimpot R1 until the connected voltmeter reads about 500 V. Please be very careful not to touch any circuit conductors while you are doing this—the results could be deadly.

The Geiger-Müller tube is moderately fussy about its operating voltage, but you don't have to get it precisely perfect. The Geiger-Müller tube demands an operating voltage between 490 and 550 V. A voltage below 490 V will not permit the Geiger-Müller tube to function. If the applied voltage is greater than 550 V, the Geiger-Müller tube will overheat, and will probably be damaged.

Once this voltage has been set, disconnect all power from the circuit, and remove the voltmeter. Connect the Geiger-Müller tube.

To test the radioactivity monitor, you will need a radioactive source. A luminous clock dial and an ionization-type smoke detector are fairly good sources. It does not have to be a highly radioactive source—in fact it shouldn't be—that would be much too dangerous. Just so the test source is more radioactive than the general background. Trimpot R7 and potentiometer R10 are interacting volume controls for the amplifier stage. Calibrate R7 to a good strong signal level, then use R10 as a front-panel volume control. In some cases, you might prefer to replace R10 with a fixed resistor, or even eliminate this component entirely (for the maximum output volume). In this last case, the bottom half of capacitor C8 would be connected directly to the speaker.

When you bring the project's probe (the Geiger-Müller tube) near the radioactive test source, you should hear a number of rapid clicks from the speaker. Adjust the volume so you can hear them clearly. Don't bother trying to count them—you won't get a meaningful or accurate measurement, however, you can get a good ballpark reading of the general level of radioactivity. The closer the probe is brought to the source, the more clicks you will hear per second, and vice versa.

Try experimenting with different types of shielding between the radioactive source and the Geiger-Müller tube. Even a few sheets of paper should reduce the number of clicks noticeably. A sheet of lead will reduce them even more.

Anywhere you expect there might be radioactive contamination, check it out with your radioactivity monitor. If you just get a few stray clicks, not much different than the ordinary background level, don't worry about it. However, if you find your monitor clicking madly away at a certain location, you might have a problem there. Try to isolate (shield) and/or remove the excessively radioactive objects as soon as possible, before your cumulative radioactive exposure level gets dangerously high. This project can literally be a life saver.

Index

A
adapting and changing projects, 1-2
air ionizer, 152-162, **155**, **156**, **158**
  high-voltage corona discharge, 158
  ion wind, 154
  ionization processes, 152-153
  operation, 156
  safety tips, 161-162
  serotonin vs. ionizaton, 154
  shielding, 155-156, **155**, **156**
air-conditioning energy saver, 84-93, **87**, **88**
air-pressure switches, 26-27
alpha radiation, 205-206
anemometer, 135
applications for electronic sensors, 1
atmosphere-related projects, 125-162
  air ionizer, 152-162, **155**, **156**, **158**
  corona discharge, 158
  ion wind, 154
  ionization process, 152-153
  operation, 156
  serotonin, 154
  shielding, 155-156, **155**, **156**
  anemometer, 135
  heater humidifier, 148-152, **149**
  humidity and moisture sensors, 125
  humidity meter, 137-148, **141**, **143**
    calibration, 146-148, **147**
    hygroscopic salts, 140
    relative humidity, 137-138
  hygrometer (*see* humidity meter)
  ionizaton sensors, 125
  relative humidity, 137-138
  wind sensors, 125
  wind-speed indicator, 125-137, **126**, **127**, **128**, **131**
    anemometer, 135
    calibration, 135-136
    timing cycle, 134-135

B
batteries, solar cells (*see* photovoltaic solar cells)
BCD counter, 74C90 IC, **79**
BCD-to-seven-segment converter, 74C47 IC, **78**
beta radiation, 206
biasing, semiconductors, 7
binary counter, 74C93 IC, **79**
bow-shock region, electromagnetic fields, 194
burned-out bulb indicator, remote, 182-185, **184**

C
Celsius temperature scale, conversion equations, 62-63
changing and adapting projects, 1-2
closed-loop system, automation thermostat/equalizer, **69**
coefficient of temperature, 2-3, 4-5
converter, BCD-to-seven-segment, 74C47 IC, **78**
cosmic radiation, 207
counters
  BCD counter, 74C90 IC, **79**
  binary counter, 74C93 IC, **79**
cross-fader, light, 171, **172**
crystals as pressure sensors, 27, **27**
current, temperature vs. current flow, 4-5

D
dark-operated relay, 165-167, **166**
dimmer, light dimmer, 169-171, **170**
diode-connected transistor, **58**
ducts, Van Allen belts, 194

E
electrochemical (gas) sensors, 30, 31
electromagnetic fields
  bow-shock region, 194
  detector, 191-200, **198**
  devices producing fields, 196
  ducts of Van Allen belts, 194
  earth as magnet, 192-193
  health risks, 190, 191-192, 196-198
  magnetosphere, 195-196
  magnetotail, 194
  naturally occurring fields, 192-193
  solar wind, 193-194, **194**
  strength of field, 197
  Van Allen belts, 194, **195**
  X-rays, 191-192
electronic nose (*see* electrochemical sensors)
energy saver for air-conditioning, 84-93, **87**, **88**
equalizer, temperature equalizer, 69-73, **69**, **71**

F
555 timer, **89**
556 dual timer, **90**
Fahrenheit temperature scale, conversion equations, 62-63
fan controller, heat-activated, 81-84, **82**
flooding alarm
  basic design, 113-116, **114**
  visual-indicator type, 116-119, **116**
frequencies of light, 163

G
gamma radiation, 207-208
gas sensors (*see* electrochemical sensors)
Geiger counter, 31, 208
Geiger-Muller tubes, 208-211
germanium diodes as temperature sensors, 6
guest-greeter light, automated, 175-179, **176**

H
Hall effect magnetic sensors, 29-31, **30**, **31**
heat sinks, temperature-sensitive components, 3
heat-leak snooper, 34-40, **36**, **39**
  alternate design, 40-44, **41**, **42**
heater humidifier, 148-152, **149**
hot-spot locator, 44-49, **46**
humidifier, heater humidifier, 148-152, **149**
humidity meter, 137-148, **141**, **143**
  calibration, 146-148, **147**
  hygroscopic salts, 140
  relative humidity, 137-138

I

ice-point bath, 57
infrared light, 163
integrated circuits
   pressure sensors, 27-28
   temperature sensors, 16
   transducers, 27-28
interleave pattern, printed circuit board, **104**
ion wind, 154-155
ionization, 152-153
ionizer (*see* air ionizer)

K

Kelvin temperature scale, conversion equations, 63-64

L

light activated silicon-controlled rectifier (LASCR), 24
light cross-fader, 171, **172**
light dimmer, 169-171, **170**
light emitting diodes (LED), 23
light sensors (*see also* light-related projects), 16-24
   LASCRs, 24
   light emitting diodes (LED), 23
   light-dependent resistor (LDR) (*see* photoresistors)
   optoisolators, 23
   photodiodes, 23, **23**
   photoelectric effect, 16
   photons and light energy, 16
   photoresistor, 22-23, **22**
      applications, 23
      light/resistance ratio, 22-23
      schematic symbol, 22, **22**
   photosensitivity, 16-17
   phototransistor, 23-24, **24**
      PNP vs. NPN, 24
      schematic symbol, 23-24, **23**
   photovoltaic solar cells, 17-22, **18**
      current detector, 19
      object detector, 19-20, **20**, **21**
      output current, 19
      output voltage, 18
      parallel configuration, 19
      schematic symbol, 20, **21**, 22
      sensors, 19
      series configuration, 19
      voltage detector, 19
light-dependent resistor (LDR) (*see* photoresistors)
light-operated relay, 164-165, **164**, **165**
light-related projects (*see also* light sensors), 163-185
   burned-out bulb indicator, remote, 182-185, **184**
   cross-fader, 171, **172**
   dimmer, light dimmer, 169-171, **170**
   frequencies of light, 163
   guest-greeter light, automated, 175-179, **176**
   infrared light, 163
   night light, self-activating, 167-169, **168**
   photosensitive automatic porch light, 179-182, **180**
   relay, dark-operated, 165-167, **166**
   relay, light-operated, 164-165, **164**, **165**
   sequential light controller, 173-175, **173**
   spectrum of light, 163
   ultraviolet light, 163
liquid sensor (*see also* liquid-related projects)
   basic design, 98-102, **99**
   probes for liquid sensor, **101**
liquid-related projects (*see also* liquid sensor), 97-124
   damage to components from liquids/water, 97
   dc power only for liquid-related projects, 98
   flooding alarm
      basic design, 113-116, **114**
      visual indicator, 116-119, **116**
   interleave pattern on printed circuit board, **104**
   liquid sensor, basic design, 98-102, **99**
   locating liquid-sensor circuits away from liquids/water, 97
   moisture detector, 102-105, **103**
   plant-watering monitor, 105-109, **106**
      deluxe design, 110-113, **110**
   probes of liquid sensor, **101**
   sump-pump controller, 119-121, **120**
   water heater controller, 121-124, **123**

M

magnetic reed switches, 29, **29**
magnetic sensor, Hall effect sensor, 29-31, **30**, **31**
magnetosphere, 195-196
magnetotail, electromagnetic fields, 194
mechanical pressure sensors, 28
mechanical switches, 24-25
mercury switches, 28-29, **28**
microphones as sensors, 31
Micro Switch (*see* snap-action switches)
moisture detector, 102-105, **103**

N

N-type semiconductors, 7
night light, self-activating, 167-169, **168**

O

Ohm's law, 4-5, 83
optoisolators, 23
over/under temperature alert, 49-52, **50**

P

P-type semiconductors, 7
photodiodes, 23, **23**
   light emitting diodes (LED), 23
photoelectric effect, 16
photons, 16
photoresistors, 22-23, **22**
   applications and uses, 23
   light/resistance ratio, 22-23
   nonpolarity, 22
   schematic symbol, 22, **22**
photosensitive automatic porch light, 179-182, **180**
photosensitivity, 16-17
phototransistors, 23-24, **24**
   NPN phototransistors, 24
   optoisolators, 23
   PNP phototransistors, 24
   schematic symbol, 23-24, **24**
photovoltaic solar cells, 17-22, **18**
   current detection using photovoltaics, 19
   object detection using photovoltaics, 19-20, **20**, **21**
   output current averages, 19
   output voltage averages, 18
   parallel configuration, 19
   schematic symbol, 20, **21**, 22
   sensor use of photovoltaics, 19
   series configuration, 19
   voltage detection using photovoltaics, 19
piezoelectric effect, 27
plant-watering monitor
   basic design, 105-109, **106**
   deluxe design, 110-113, **110**
   probes, **108**
PN junction in semiconductor, temperature sensors, 7, **8**
porch light, photosensitive, 179-182, **180**
position switches, 28-29
   magnetic reed switches, 29, **29**

hygrometer (*see* humidity meter)
hygroscopic salts, humidity meter, 140

Index

mercury switches, 28-29, **28**
tilt switches (*see* mercury switches)
power-usage meter, 93-96, **93**
pressure sensors, 24-28
  air-pressure switches, 26-27
  crystals as pressure sensors, 27, **27**
  hybrid piezoresistive IC pressure transducers, 27-28
  mechanical pressure sensors, 28
  mechanical switches, 24-25
  Micro Switches (*see* snap-action switches)
  piezoelectric effect, 27
  pressure switch, 25
  snap-action switches, 25-26, **26**
pressure switches, 25, **25**
printed circuit board, interleave pattern, **104**
pump controller, sump-pump, 119-121, **120**

R

radiation and radioactivity, 187-188, 201-211
  alpha radiation, 205-206
  beta radiation, 206
  cosmic radiation, 207
  gamma radiation, 207-208
  Geiger counter, 208
  Geiger-Muller tubes, 208-211
  health risks, 201-205
  monitor, 201-211, 201
  naturally occurring radiation, 204, 207-208
  radon gas, 204-205
  sensors, 31
  X-rays, 204, 207
radon gas, 204-205
Rankine temperature scale, conversion equations, 64
relative humidity, 137-138
relays
  cascaded relays, **68**
  dark-operated, 165-167, **166**
  light-operated, 164-165, **164**, **165**
  parallel connected relays, **90**
remote burned-out bulb indicator, 182-185, **184**
resistance
  parallel configuration, adding resistance, 14-15
  series configuration, adding resistance, 13
  temperature coefficients, 2-3, 4-5

S

74C47 BCD to seven-segment converter, **78**
74C90 BCD counter, **79**
74C93 binary counter, **79**
7555 timer, **89**
Seebeck effect, 6
self-activating night light, 167-169, **168**
semiconductors
  biasing, forward vs. reverse, 7
  N-type, 7
  P-type, 7
  photosensitivity, 16-17
  PN junctions, 7, **8**
  temperature-sensitivity, 3
  thermal runaway, 5
sensors, 1-33
  applications for electronic sensors, 1
  changing and adapting projects, 1-2
  electrochemical (gas) sensors, 30, 31
  electronic sensors, 1
  exotic sensors, 29-32
  Hall effect magnetic sensors, 29-31, **30**, **31**
  homemade sensors, 32
  light sensors, 16-24
  microphones as sensors, 31
  piezoelectric effect, 27
  position switches, 28-29
  pressure sensors, 24-28
  radioactivity sensors, Geiger counters, 31
  temperature sensors, 2-16
sequential light controller, 173-175, **173**
serotonin and ionization of air, 154
silicon diodes as temperature sensors, 6-8
snap-acting switches, 25-26, **26**
solar batteries (*see* photovoltaic solar cells)
solar cells (*see* photovoltaic solar cells)
solar wind, electromagnetic fields, 193-194, **194**
spectrum of light, 163
sump-pump controller, 119-121, **120**
superconductivity, temperature and heat energy, 4
switches
  air-pressure switches, 26-27
  magnetic reed switches, 29, **29**
  mechanical switches, 24-25
  mercury switches, 28-29, **28**
  Micro Switch (*see* snap-action switches)
  position switches, 28-29
  pressure switches, 25, **25**
  snap-action switches, 25-26, **26**
  tilt switches (*see* mercury switches)

T

technological environmental risks, 186-211
acquired immune-deficiency syndrome (AIDS) example, 188
electromagnetic fields
  bow-shock regions, 194
  detector, 191-200, **198**
  devices producing fields, 196
  ducts, Van Allen belts, 194
  earth as magnet, 192-193
  health risks, 190, 191, 196-198
  magnetosphere, 195-196
  magnetotail, 194
  naturally occurring fields, 192-193
  solar wind, 193-194, **194**
  strength of field, 197
  Van Allen belts, 194, **195**
proof vs. disproof, 187-190
radioactivity, 187-188
  alpha radiation, 205-206
  beta radiation, 206
  cosmic radiation, 207
  gamma radiation, 207-208
  Geiger counter, 208
  Geiger-Muller tubes, 208-211
  health risks, 201-205
  monitor, 201-211
  naturally occurring radiation, 204, 207-208
  radon gas, 204-205
  X-rays, 204, 207
technophobes vs. technophiles, 186
X-rays, health risks, 191-192
temperature coefficients, 2-3, 4-5
temperature sensors (*see also* temperature-related projects), 2-16
  coefficient of temperature, positive and negative, 2-3, 4-5
  current flow vs. temperature changes, 4-5
  dedicated temperature sensors, 5-16
  germanium diodes as temperature sensors, 6
  heat sinks and temperature-sensitive components, 3
  integrated circuits as temperature sensors, 16
  resistance vs. temperature changes, 2-3, 4-5
  Seebeck effect, 6
  semiconductors

Index

temperature sensors continued,
  biasing, forward vs. reverse, 7
  PN junctions, 7
  temperature-sensitivity, 3
  sensitivity of components to temperature, 3-4
  silicon diodes as temperature sensors, 6-8
  superconductivity and heat energy, 4
  thermal runaway, 5
  thermistors, 9-15
    effective resistance, 13, 15
    negative temperature coefficient, 12
    nonlinearity, 12
    parallel configuration, 14-15
    polarity: nonpolarity of thermistor, 9
    positive temperature coefficient, 12
    range-resistor, 13, **13**, **14**, **15**
    resistance alteration, 13
    resistance calculation, 13
    schematic symbols, 9-11, **10**, **11**, **12**
    sensors, 12
    series configuration, 13
  thermocouples, 6
  thermostats, 6
  tolerance ratings vs. temperature sensitivity, 2-3
  transistor as temperature sensors, 8, **9**
  voltage vs. temperature changes, 4-5
  zener diode as temperature sensors, 8
temperature-related projects (*see also* temperature sensors), 34-96
  air-conditioning energy saver, 84-93, **87**, **88**
  Celsius vs. Fahrenheit conversion equations, 62-63
  equalizer, temperature equalizer, 69-73, **69**, **71**
  fan controller, heat-activated, 81-84, **82**
  heat-leak snooper, 34-40, **36**, **39**
    alternate design, 40-44, **41**, **42**
  hot-spot locator, 44-49, **46**

Kelvin temperature scale conversion, 63-64
  over/under temperature alert, 49-52, **50**
  power-usage meter, 93-96, **93**
  Rankine temperature scale conversion, 64
  thermometer, electronic adaptor, 58-65, **60**
    simple design, 52-58, **53**, **55**, **58**
  thermometer, long-term, 73-81, **74**, **75**
  thermostat, electronic, 65-69, **66**
thermal runaway, 5
thermistors, 9-15
  effective resistance calculation, 13, 15
  negative temperature coefficient thermistors, 12
  nonlinearity effects, 12-13
  parallel configuration, resistances, 14-15
  polarity: nonpolarity of thermistors, 9
  positive temperature coefficient thermistors, 12
  range-resistor added to thermistor, 13, **13**, **14**, **15**
  resistance alteration, 13
  resistance calculations, 13-14
  schematic symbols, 9-11, **10**, **11**, **12**
  sensors, 12
  series configuration, resistances, 13
thermocouples, 6
  Seebeck effect, 6
thermometer, electronic adaptor, 58-65, **60**
  Celsius temperature scale conversions, 62-63
  Fahrenheit temperature scale conversions, 62-63
  ice-point bath, 57
  Kelvin temperature scale conversion, 63-64
  Rankine temperature scale conversion, 64
  simple design, 52-58, **53**, **55**, **58**
thermometer, long-term, 73-81, **74**, **75**

thermostats, 6
  electronic thermostat project, 65-69, **66**
  equalizer, temperature equalizer, 69-73, **69**, **71**
tilt switch (*see* mercury switches)
timers
  555 timer, **89**
  556 dual timer, **90**
  7555 timer, **89**
transducers as pressure sensors, 27-28
transistors
  diode-connected transistor, 58, **58**
  temperature changes vs. performance, 4-5
  temperature sensor use, 8, **9**
  thermal runaway, 5
twisted pair, **58**

U

ultraviolet light, 163

V

Van Allen belts, 194, **195**
voltage, temperature vs. voltage, 4-5
voltage-divider network, **55**

W

water heater controller, 121-124, **123**
weather sensors (*see* atmosphere-related projects)
wind-speed indicator, 125-137, **126**, **127**, **128**, **131**
  anemometer, 135
  calibration, 135-136
  timing cycle, 134-135

X

Xrays, 204, 207
  health risks, 191-192

Z

zener diode as temperature sensors, 8